Bamboo Flowering and Rodent Control

The Editors

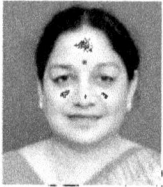

Dr. Veena Tandon FNASc, MSc and PhD from Panjab University is a Professor of Zoology, serving at North-Eastern Hill University, Shillong for the last >30 years as a teacher and researcher of Parasitology. She has studied in depth the worm parasite spectrum in vertebrates of economic and food value in Northeast India, with emphasis on geographical distribution and influence of environmental factors on prevalence and transmission of parasitic infections in the region. She is recipient of the coveted 'E.K. JanakiAmmal National Award in Animal Taxonomy' (2006), conferred by Ministry of Environment & Forests, Government of India.

Dr. S.K. Barik is Professor of Ecology at the Department of Botany, North-Eastern Hill University, Shillong, and is currently working as the Head of the Department. He is an accomplished ecologist with twenty six years of research experience in the field of Ecology and Natural Resources Management. He is pursuing frontline research in diverse areas of ecology including ecosystem structure and function, ecological modelling, and population and molecular ecology. He served/ is serving as Chairman/Member of several Expert Committees/Task Forces constituted by Ministry of Environment and Forests, Planning Commission, and Ministry of Science and Technology, Govt. of India. He has published eleven books and 70 research papers. He is Member of the Editorial Boards of *Tropical Ecology* and *Journal of Biodiversity*. He is a LEAD Fellow, awarded by Leadership in Environment and Development (LEAD) International, Rockefeller Foundation, USA.

Bamboo Flowering and Rodent Control

— Editors —

Veena Tandon
Saroj K. Barik

Department of Science & Technology
Ministry of Science & Technology
Governemnt of India
New Delhi

2015

Regency Publications
A Division of
Astral International Pvt. Ltd.
New Delhi – 110 002

Cataloging in Publication Data--DK
Courtesy: D.K. Agencies (P) Ltd. <docinfo@dkagencies.com>

Workshop on 'Bamboo Flowering and Rodent Control' (2010 : Guwahati, India)
Bamboo flowering and rodent control / editors, Veena Tandon, Saroj K. Barik.
   pages cm
Includes bibliographical references and index.
  ISBN 9789351306214 (International Edition)

    1. Bamboo--Taiwan--Flowering--Congresses. 2. Bamboo--India, Northeastern--Flowering--Congresses. 3. Rodents--Taiwan--Control--Congresses. 4. Rodents--India--Control--Congresses.   I. Tandon, Veena, editor. II. Barik, S. K., editor. III. North Eastern Hill University.organizer. IV. Title.

    DDC 632.69350951249   23

Published by   : **Regency Publications**
                     A Division of
                     Astral International Pvt. Ltd.
                     – ISO 9001:2008 Certified Company –
                     4760-61/23, Ansari Road, Darya Ganj
                     New Delhi-110 002
                     Ph. 011-43549197, 23278134
                     E-mail: info@astralint.com
                     Website: www.astralint.com

Laser Typesetting   : **Classic Computer Services**, Delhi - 110 035

Printed at   : **Thomson Press India Limited**

PRINTED IN INDIA

Foreword

The phenomenon of bamboo flowering and associated problems are of grave concern to Asia-Pacific region. Substantial bamboo forests in the region have disappeared due to gregarious flowering of the respective species. In North-East India, large tracks of bamboo forests comprising of *Melocanna baccifera* flowered during 2006-2010 and sporadic flowering of *Dendrocalamus hamiltonii* took place during the past 15 years. However, the science of bamboo flowering, particularly the mechanism of flowering and the subsequent death and associated increase in rodent populations by and large remain an enigma. In view of bamboo being an important bioresource contributing to economies of many countries in the developing world, a workshop on 'Bamboo Flowering and Rodent Control' was organized by North-Eastern Hill University, Shillong during November 11-13, 2010 under the joint initiative of Department of Science and Technology (DST), Government of India and the National Science Council (NSC), Taiwan.

The present compendium, titled, "Bamboo Flowering and Rodent Control", is an outcome of deliberations at this workshop. Seven experts from Taiwan and five from India have contributed to this volume presenting the current scenario of bamboo research, with special emphasis on bamboo flowering and rodent control, in both the countries. The papers included in this volume range from taxonomy of bamboo, the plausible mechanisms of bamboo flowering using molecular tools, predictive modelling of bamboo flowering and rodent control measures. I congratulate the editors, Prof. Veena Tandon and Prof. S.K. Barik, for bringing out this publication, which is timely and highly relevant to the problems of bamboo flowering and associated growth of rodent populations. I am sure, the compendium will be a very good reference material for researchers in this interesting but less studied discipline.

Pramod Tandon
Former Vice Chancellor,
North-Eastern Hill University,
Shillong – 793 022

Preface

Bamboo that belongs to the family Poaceae and sub-family Bambusoidae plays an integral role in the social, cultural and economic life of the people in Asian countries. Bamboo contributes immensely to the rural livelihood in these countries. The industrial use of bamboo in the manufacture of paper pulp dates back to several centuries. In the recent past, a number of bamboo-based wood substitutes have been developed and the industrial applications of bamboo are growing rapidly contributing significantly to the economy of the bamboo- growing nations.

The factors that cause flowering in bamboos remain a botanical mystery. Many causes have been attributed to trigger gregarious flowering in bamboo but it has not been possible to establish their consistency. For instance, environmental factors do influence the growth of plants, and trigger flowering in majority of them. However, the eco-physiological factors (*e.g.* drought, nutrient deficiency, solar radiation and photo-periodism) that may be consistently responsible for bamboo flowering have not been identified yet. The molecular mechanism that triggers flowering in bamboo is still being investigated. There is no scientific method yet developed for predicting flowering in bamboo. Most bamboo species suddenly flower either gregariously or sporadically, and ultimately die at the end of it. Such death is attributed to reproductive exhaustion caused by the movement of food reserves from the vegetative parts to the reproductive parts. It takes substantial time for the new generation to establish and restore back the original bamboo forest.

Flowering in bamboo is believed to be a bad omen in many cultures. People believe that famine, death and natural disasters are associated events of bamboo flowering. For example, in Mizoram, a north-eastern state of India, flowering in *Bambusa tulda* (Mizo name *Rawthing*) is associated with the 'Thingtam' famine and devastation. The 'Mautam' famine is associated with *Melocana baccifera* (Mizo name *Mautak*). 'Mautam' is said to be more severe as this species bear large fruits and can thus support a larger rat population. These famines have occurred in years associated with the gregarious flowering of bamboos. Both the species are reported to flower every 48 years but not at the same time. *B. tulda* flowers 18 years after *M. baccifera*. *B. tulda* gregariously flowered in the years1880 through 1884 and in 1928 through 1929. Flowering in *M. baccifera* alsotook place in 1910-1912 and 1958-1959 in northern India. In fact, when bamboo flowers gregariously, it leads to economic, social and ecological disaster. During gregarious flowering, large quantities of seeds are produced. Many animals, particularly rodents feed on these nutritious seeds and reproduce fast to increase their population sizes. Once the seeds are exhausted due to germination, the rats/rodents depend on crops in the vicinity, thus causing famine.

In north-eastern India, *M. baccifera* flowered recently during 2004-2010 in different parts of the region. Similarly, another bamboo species, *Dendrocalamus hamiltonii*, flowered both gregariously and sporadically from 1996-2010. Once the bamboos died after flowering, the people whose livelihood was dependent on the bamboo resources were severely affected. Besides, the dead culms caught fire easily during dry winter months destroying substantial forest areas in the respective states. The forests were also opened up providing habitats for invasion by invasive alien species causing serious long-term damage to forest environment. Recent death of large populations of *M. baccifera* and *D.hamiltonii* spread over large tracts of forest areas in the region has been a cause of concern owing to the ecological, social and economic crises that set forth following bamboo flowering.

In view of this, the Department of Science and Technology (DST), Government of India sponsored a workshop under Indo-Taiwan joint initiative to brain storm the scientific, technological, social and environmental issues related to bamboo flowering, and the associated problem of rodent population growth. Since the problem is common to the entire Asia-pacific region, the participation of scientists both from India and Taiwan was of mutual interest. The National Science Council (NSC), Taiwan was the collaborator of this programme. The workshop was organized and hosted by North-Eastern Hill University, Shillong, India and was held at Guwahati during November 11-13, 2010. This book is an outcome of this 3-days workshop. It has included 12 selected papers presented during the workshop. The editors thank all the authors who contributed to this compendium. The editors also acknowledge the financial support received from DST, Government of India for this initiative.

We hope, this publication will be a good reference source for those who are interested in bamboo research in general, and flowering of bamboos and rodent control, in particular.

Veena Tandon

S.K. Barik

Editors

Contents

Chapter 1
Bamboo Resources, Management and Researches in Taiwan

Dar-Hsiung Wang

Taiwan Forestry Research Institute, Taiwan,
E-mail: dhwang@tfri.gov.tw

Introduction

Bamboo belongs to the family *Gramineae* and has about 90 genera with over 1 200 species around the world (Lobovikov *et al.*, 2007), of which 14 genera and 120 species are found in Asia (Azmy *et al.*, 1997). Bamboo flowers rarely and in irregular cycles, which are not yet clearly understood. Bamboo is naturally distributed in the tropical and subtropical belt between approximately 46° north and 47° south latitude, and is commonly found in Africa, Asia, and Central and South America. Some species may also grow successfully in mild temperate zones in Europe and North America. Bamboo is an extremely diverse plant, which easily adapts to different climatic and soil conditions. Dwarf bamboo species grow to only a few centimeter (cm), while medium-sized bamboo species may reach a few meters (m) and giant bamboo species grow to about 30 m, with a diameter of up to 30 cm. Bamboo stems are generally hard and vigorous, and the plant can survive and recover after severe calamities, catastrophes and damage.

Due to rapid growth in the use of bamboo, the concern about the sustainability of global bamboo resources becomes eminent. Despite the successful bamboo trade, very little is known about the actual status and dynamics of the global bamboo resource base. One of the first attempts to assess bamboo resources on a global scale was carried out by FAO and the United Nations Environment Program in 1980 covering 13 countries known to possess substantial bamboo resources as part of the Global Forest Resources Assessment 1980 (FRA 1980).

Despite the importance of bamboo worldwide, global statistics on its resource base, production and trade remain rather scarce and inconsistent. Absence of reliable and comprehensive data on bamboo resources and utilization has been limiting their potential to contribute to poverty reduction. In the past, both FAO and INBAR (International Network for Bamboo and Rattan), under their respective mandates, have addressed the issue of bamboo resources assessment through various activities and studies. Recently, a thematic study on bamboo was developed by FAO and INBAR jointly in the framework of FAO's *Global Forest Resources Assessment 2005* (FRA 2005), with the aim of filling the gap in global information and providing a first, comprehensive assessment of the world's bamboo resources (Lobovikov *et al.*, 2007).

Based on data reported to FAO/INBAR by countries around the world in 2005, Asia owns the richest bamboo resources with 65 per cent of total world's bamboo resources, followed by America (28 per cent) and Africa (7 per cent). In Asia, the major bamboo producing countries are India (almost 11.4 million hectares) and China (over 5.4 million hectares), followed by Indonesia (2 million hectares) and the Lao People's Democratic Republic (1.6 million hectares). India accounts for roughly half the total area of bamboo reported for Asia and, together with China, approximately 70 percent. Over the last 15 years, the bamboo area in Asia has increased by 10 percent, primarily due to large-scale planting of bamboo in China and, to a lesser extent, in India.(Lobovikov 2007). Compared to the total forest area in the reporting countries, the proportion of bamboo forest is about 3.2 per cent.

Bamboo Species in Taiwan

Since Taiwan is sitauted on the Tropic of Cancer and high elevation ones are present in the central mountain range in Taiwan, different types of bamboos are found in Taiwan. There are 46 species of bamboo found in Taiwan, of which 20 are indigenous and 26 exotic (Lin 1967). Bamboo shoots and culms grow from the dense rhizome system. There are two main categories of rhizomes: monopodial and sympodial. Monopodial type rhizomes grow horizontally, often at a surprising rate, and thus their nickname of 'runners'. The rhizome buds develop either upward, generating a culm, or horizontally, with a new tract of the rhizomal net. *Phyllostachys* and *Teragonocalamus* are two genera of monopodial type rhizomes found in Taiwan. The common species of monopodial bamboo in Taiwan include *Phyllostchys makino* and *Phyllostachys pubescnes*. Monopodial type bamboos generate an open clump with significant culm distance between the two consecutive culms can be invasive.

Sympodial bamboos are short and thick, and the culms above ground together form a compact clump, which expands evenly around its circumference. Three genera of Sympodial type rhizomes are found in Taiwan. They are *Bambusa, Dendrocalamus* and *Schizostachyum*. The major species of sympodial type rhizomes in Taiwan are *Dendrocalamus latiflorus, Dendrocalamus giganteus, Bambusa oldham, Bambusa dolichoclada* and *Bambusa Stenostachya.* Because of the compact clump, they are not invasive.

In addition to these two traditional types of bamboo rhizomes, Lin (1976), identified an intermediate type rhizomes. In Taiwan the example of intermediate type is *Melocanna baccifera* (Lü 1996).

Bamboo Resources in Taiwan

Bamboo is an integral part of forestry, but it is also widely spread outside forests, including farmlands, riverbanks, roadsides and urban areas in Taiwan.

Table 1.1: Area and Stocking of Bamboo in Taiwan (year 2000).

Hsien/City	Area(ha)	Stocking(thousand culms)
Taipei City	565	3,655
Kaoshiung City	109	317
Taipei Region (Northern)	12,913	159,774
Taipei Hsien	9,548	108,251
Iland Hsien	2,648	40,369
Keelung City	717	11,153
Hsinchu Region	31,444	466,361
Taoyuan Hsien	6,519	97,107
Hsinchu Hsien	9,591	141,359
Miaoli Hsien	15,334	227,895
Hsinchu City		
Taichung Region (Central)	29,393	115,681
Taichung Hsien	436	128,301
Changhua Hsien	1,171	7,598
Nantou Hsien	22,310	69,720
Taichung City	1,551	10,062
Tainan Region (Southern)	44,907	132,502
Yunlin Hsien	7,033	27,835
ChiayiHsien	21,713	47,270
Tainan Hsien	16,074	57,209
Chiay City	87	188
Tainan City		
Kaoshiung Region (Southern)	16,236	50,706
Kaoshiung Hsien	15,319	48,039
Pingtung Hsien	918	2,667
Taitung Region (East)Taitung Hsien	5,350	39,595
Haulien Region (East)(Haulien Hsien)	8,598	140,330
Ponghu Region (Ponghu Hsien)		
Total	**149,516**	**1,108,921**

Source. Lü (2001).

The total area of bamboo resources in Taiwan was 75,275 ha in 1962, 175,638 ha in 1971, and 149,516 ha in 2000(Lin *et al.,* 1962, Lü 2001). Table 1.1 displays the growing stock of bamboo in Taiwan in 2000. From Table 1.1, it reveals that the area of bamboo in Tainan region ranks the first, followed by Hsinchu region, and Taichung region. Based on administration unit, the bamboo area in Nantou Hsien is the largest one, second in Chiayi Hsien, and followed by Tainan Hsien. In Taiwan, except Hsinchu City, Tainan City and Ponghu Hsien, bamboo resources can be found in all other places.

Taiwan Forestry Research Institute, in cooperation with JCRR (Joint Commission on Rural Reconstruction), carried out an island-wide bamboo resources survey from Oct. 1960 to Dec.1961 (Lin *et al.,* 1962). In this survey, information on distribution, growing stock and annual yield per hectare for six principal bamboo species were collected. The acreage of six principal bamboo species in Taiwan is shown in Table 1.2. It indicates that *Phyllostchys makinoi* is the most abundant species with acreage of 52.53 per cent of total bamboo forests in the island. Moreover, this survey showed the location of habitats of the dominant species in bamboo forests. *Phyllostchys makino, Dendrocalamus latiflorus,* and *Phyllostachys pubescens,* for example, are the dominant species in Central Taiwan, *Phyllostchys makinoi* and *Bambusa oldhami* are in Northern Taiwan, and *Bambusa Stenostachya, Dendrocalamus latiflorus,* and *Bambusa dolichoclada* are in Southern Taiwan (Lin *et al.,* 1962), respectively.

Table 1.2: Area of Bamboo in Taiwan.

Species	Area (ha)	Percentage (per cent)	Endemic/Exotic
Phyllostachys makinoi	39,542	52.53	Endemic
Phyllostachys pubescens	2,297	3.05	Exotic
Dendrocalamus latiflorus	18,146	24.11	Exotic
Bamausa stenostachya	8,011	10.64	Exotic
Bambusa dolichoclada	3,706	4.92	Endemic
Bambusa oldhami	3,573	4.75	Exotic
Total	**75,275**	**100**	

Source: Lin *et al.* (1962).

Almost ten years later, another island-wide survey of bamboo resources was carried out during the period from Jan.1971 to June 1972 (Dai *et al.,* 1973). In this survey, the aerial photographs were used to estimate bamboo resources around Taiwan. In contrast with the previous survey, mixture stands of bamboo with trees or crops were included in this survey. Table 1.3 illustrates the results of this survey. A comparison with Table 1.2 reveals a big change in acreage distribution among dominant species. For example, *Dendrocalamus latiflorus,* owing to having a large area of mixture with trees or crops, occupied more than half of the total bamboo forests. Most dominant species of bamboo in Taiwan, such as *Dendrocalamus latiflorus, Bambusa Stenostachya, Bambusa oldhami* and *Phyllostachys pubescens* are the exotic, therefore, the management of the exotic bamboo is quite important in Taiwan.

Table 1.3: Area under different Bamboo Species in Taiwan.

Species	Area in Pure Bamboo (ha)	Area in Bamboo Mixed Tree/Crop (ha)	Total (ha)	Percentage (per cent)
Dendrocalamus latiflorus	14,911	75,954	90,865	51.73
Bamausa stenostachya	17,280	13,378	30,658	17.46
Bambusa oldhami	4,434	25	4,459	2.54
Phyllostachys edulis	3,296		3,296	1.88
Phyllostachys makinoi	43,774		43,774	24.92
Other	735	1,851	2,586	1.47
Total	84,430	91,208	175,638	100

Source: Dai *et al.* (1973).

Bamboo Distribution in Taiwan

In Taiwan, bamboo can be found on high mountains, hill slopes, river banks, logged-over areas and on flat lands. While Taiwan is a small island (*i.e.*, 3,600 km² area), due to the heterogeneous geography (4000 meters range in elevation), bamboos are distributed across four climatic zones, with most abundant in sub-tropical and tropical zone followed by temperate zone, and a few were found in alpine zone. Compared to Japan and South-eastern countries in Asia, Taiwan is unique to have bamboo distribution over four climatic zones.

Table 1.4 shows the major bamboo species distributed along an elerstional gradient in Taiwan. Usually, monopodial bamboos are found in temperate regions, and sympodial bamboos are found in sub-tropical and tropical regions. The vegetation of bamboo can be pure stand or mixed with other tree species in the forest.

Table 1.4: Distribution of Bamboo Species by Elevation in Taiwan.

Elevation Above Sea Level (Meters)	Bamboo Species
Above 3000	*Yuahania niitakayamensis*
2000-3000	*Yuahania niitakayamensis*
1000-2000	*Phyllostachys makinoi*
	Phyllostachys pubesens
	Sinobambusa kunishii
500-1000	*Bambusa dolichoclada*
	Sinobambusa kunishii
0-500	*Bambusa dolichoclada*
	Bambusa oldhami
	Bamausa stenostachya

Bamboo Research in Taiwan

Bamboo research in Taiwan can be traced back to late in 1950. Studies included bambusaceae classification (Lin 1961,1976), propagation (Lin 1962, 1964, Lu *et al.,* 1982, Kao *et al.,* 1989), resource inventory (Lin *et al.,* 1962, Lin 1967), morphology of bamboo flowers (Lin 1974), disease (Lin *et al.,* 1979, Lin *et al.,* 1981), growth (Lin 1958, Chen *et al.,* 2009), bamboo shoot and biomass production (Kiang *et al.,* 1976, Liu and Kao 1988, Lu and Chen 1992), property and utilization (Lin *et al.,* 1976, Lin *et al.,* 1977). In additional to these, recently, bamboo research has been extended to investigate the bamboo root system, bamboo mechanical behaviors and the potential anti-disaster function of bamboo forests on landslide (Lin *et al.,* 2007, Lin *et al.,* 2009); hydrological characteristics of bamboo forest (Lu *et al.,* 2007). Moreover, the effect of management activity (thinning) on growth and biomass production in bamboo forest has been evaluated as well (Chung *et al.,* 2010).

Compared with woody species, a more rapid growth in height occurs on bamboo. Table 1.5 shows the growth rate (height) for two endemic bamboo species. It only took 72 days to complete the height growth for *Dendrocalamus latiflorus,* however, the period for height growth reduced to 29 days only for *Phyllostchys makinoi.* The height growth pattern of *Dendrocalamus latiflorus* was reflected by a tendency of a slow initial growth in the beginning, with the increasing growth rate occurred after 10 days, then it reaches the most rapid level 40 days later, and finally, the growth rate begins to reduce from 55 days. The same pattern was observed in *Phyllostchys makino* with the highest growth reaching in 10 days after the shoot initiation (Lin 1958).

Table 1.5: Height Growth for Two Endemic Bamboo Species in Taiwan.

Species	Growth Completed in Days	Increment (cm/day)
Dendrocalamus latiflorus	72	15.4±0.4
Phyllostchys makinoi	29	27.9±1.9

Source: Lin (1958).

Individuals of Bamboo are regenerated via rhizome development. The growth of culms depends on the quality of rhizome. Ueda (1960, 1963) mentioned that the productivity is highest at the age of 2-6 years for lateral monopodial culms bamboos. Depending on factors such as species, site conditions (soil and climate), and management, the number of bamboo clumps per ha and clump size varied in different locations in Taiwan. The survey of *Phyllostchys makinoi* in Tung-Tou area in Taiwan indicates that the stand density can reach as high as 10,000 stems per hectare, of which only 16 per cent of stems are in age of 1-2 years, and the mean diameter of culms was cover in young bamboos then the old (Lu and Chen 1992). Since no cutting treatment was practiced in old *Phyllostchys makinoi* bamboo forest, it caused not only reduction in the amount of rhizome, but also in lowering productivity of the forest.

In the past, bamboo constituted significantly to Taiwan's economic development. Bamboo served innumerable economic uses in Taiwan. The uses include food, paper-making, handicraft, utensils, shelter and other kinds of structures. Due to the flourishing prospects of bamboo industry, development and intensive management of bamboo forests including introduction of exotic species and improved cultivation practices were undertaken widely around Taiwan (Lin 1962). In 1960-1970, under the sponsorship of Rural Development Council, vigorous activities on bamboo research, bamboo management techniques and establishment of bamboo shoot production area were carried out around the island. However, since 1980, due to increase in labor cost and the shrinkage of bamboo industry, willingness to manage bamboo forests has reduced among the farmers. Recently, due to the big drop in culms usage of *Phyllostchys makinoi* in Taiwan, the acreage of unmanaged bamboos was increased dramatically (Watanabe 2005, *Lü 2000)*. The intermingling of old, dead, and fallen culms within the unmanaged bamboo forest resulted in the decay of the entire old stands. However, in monopodial culms type bamboos, the rhizome of individuals at the rim of stand still can be expanded outwards at this stage.

Two types of cutting systems are used in Taiwan. For *Bamausa stenostachya* clear-cutting (with 6-8 yr rotation) is generally practiced. For other five key species, selective-cutting is performed with rate of annual cut approximately one-fourth to one-third of the growing stock (Lin 1962).

In *Phyllostachys pubesens* in China, no bamboo flowering occurred for the intensive managed bamboo stands for over 150 years. On the contrary, the unmanaged bamboo stands may cause the early flowering due to the weak condition in sustenance growth (Lu 2001).

Bamboo is a plant which has not only high value in economic return but also has high benefits in terms of ecological function. In the greatly populated developing countries such as China and India, hundred millions of people earn income from bamboo. Based on the management objectives, bamboo resource management is generally classified into two types:

1. Economic benefits oriented bamboo stands with the maximum income from culms usage, bamboo shoots, and landscape aesthetics.
2. Ecological benefits oriented bamboo stands with the maximum protection services against adverse environment and weather conditions.

Bamboo Management in Taiwan

Bamboo cultivation is different depending on the type of bamboo. Five methods are widely used for the sympodial bamboos. They are sprout cutting, level transplant, vertical transplant, branch transplant, pressured mold, and seedling. Three methods are widely used for the monopodial bamboos. They are rhizome plantation with mother culms, rhizome plantation without mother culms, and seedling. Site for plantation is determined depending on bamboo type and terrain locations. Usually, sympodial bamboos are suitable to be planted at low-elevation area. The suitable area for monopodial bamboos, for instance, is 500~1600 m elevation for *Phyllostachys pubesens*, and 100~1500m for *Phyllostachys makinoi*.

Site preparation is necessary for bamboo planting. Usually, three methods including entire reclaim, stripe reclaim, patch reclaim can be used for the site preparation before bamboo planting. Planting density for sympodial large size bamboo is about 300-625 stems/ha, 500-1100 stems/ha for sympodial medium size bamboo and 278-500 stems/ha for monopodial bamboos (Lu 2001).

Bamboo tending is quite important to ensure bamboo growth. The intensive management of bamboo can not only enhance the bamboo production (*e.g.*, culms, bamboo shoots) but also delay the time for bamboo flowering because of the inhibition on reproduction growth (Jiang 2002).

Bamboo tending includes irrigation, weed control, fertilization, and harvest. The methods used for bamboo tendering vary among bamboo types and production purposes (*i.e.* bamboo shoots, bamboo culms, garden view, environmental protection). Usually, for bamboo shoots production bamboo forests, intensive cultures including hill up, irrigation, fertilization, mold and harvest are applied to increase the bamboo shoots production (Lu. 2001).

Density control is quite important to monopodial bamboos because of the emergence of new culms every year. Literature shows that for monopodial bamboos with very high density, the growth of new culms becomes worse, therefore, the productivity of bamboos will be decreased (Kao *et al.*, 1989). Therefore, harvest of old culms should be carried out each year. The amount of harvest depends on the site, species and management objective. Usually, if bamboos are planted mainly for bamboo shoots production, the density of bamboo should be low to allow the more sunlight to produce more bamboo shoots (Table 1.6). However, for culms production mainly, the density should be kept in high to produce more bamboo culms (Lu. 2001).

Table 1.6: The Density for *Phyllostachys makinoi* and *Phyllostachys pubesens* with Alternative Objectives.

Management Objectives	Bamboo Species			
	Phyllostachys pubesens		*Phyllostachys makinoi*	
	Number of Culms/ha	Number of Culms/ha by Five Age Classes	Number of Culms/ha	Number of Culms/ha by Five Age Classes
Culms mainly	6000-7500	1200-1500	13000-15000	2600-3000
Bamboo shoots mainly	3500-5000	700-1000	6000-8000	1200-1600
Both	5000-6000	1000-1200	8000-10000	1600-2000

Source. Lü. 2001.

Bamboo Utilization in Taiwan

Bamboo is a major non-wood forest product and wood substitute. Globally, it is found in all regions of the world and plays an important economic and cultural role. In addition to be used for housing, crafts, pulp, paper, panels, boards, veneer, flooring, roofing, fabrics, oil, gas and charcoal (for fuel and as an excellent natural absorbent), it is also a healthy vegetable (the bamboo shoot). Bamboo industries are now thriving in Asia and are quickly spreading across the continents to Africa and America (Lobovikov *et al.*, 2007).

Being a multipurpose plant, bamboo can be used in various forms, from traditional uses to commercial products. Bamboo has innumerable economic uses in Taiwan. In the past, most of bamboo commercial products in Taiwan were baskets, chopsticks, toothpicks, bamboo shoots, skewers, blinds, joss stick, papers, and handicraft items. Nowadays, the processing of bamboo in Taiwan is shifting from low-end crafts and utensils to high-end, value-added commodities such as laminated panels, boards, pulp, paper, mats, prefabricated houses, cloth and artistic carvings.

In addition to benefits mentioned above, the fiber of bamboo is quite suitable to pulp making with good quality and for bio-energy biomass (*e.g.*, cellulosic ethanol) production, therefore, bamboo is a good substitute for wood with shrinkage in the wood pulp. However, due to the limitations of supply of bamboo culms produced in unmanaged bamboo stands, raw material availability is a problem faced by the bamboo industry of Taiwan.

Prospect of Sustainable Development of Bamboo in Taiwan

Being a multi-purpose plant resource, bamboo is drawing an ever-increasing attention from different countries of the world including Taiwan and various international organizations. In the past decades, advances have been achieved in scientific research, production, processing and trade in Taiwan bamboo industry.

With the outstanding characteristics of short rotation, high economic value in bamboo production and advantage for sustainable management, Bamboo is the most important non-wood forest product, and is of great commercial value and has enormous potential for development in the future in Taiwan. Moreover, bamboo is an excellent garden plant, representing a unique value for visual appreciation and arts in nowadays.

Under the prerequisite that sustainable management is essential to maintain bamboo resources for numerous reasons and benefits in Taiwan, a process of integration of activities, coordination, and support to strategic policy and appropriate research and development programs need to be undertaken. It is anticipated to enhance the interests of the producers and consumers of bamboo, to upgrade the capability and technical level of bamboo development institutions and services organizations at national level and to strengthen and cooperation among different countries and regions.

In view of the great benefits associated with bamboo forests, the importance of bamboo should be realized and intensive management should be taken up for the sustainable bamboo resources management in Taiwan.

References

Azmy, H. M., Norini, H., and Wan Razali, W. M. (1997). Management guidelines and economics of natural bamboo stands. FRIM Technical Information Handbook No.15. Forest Research Institute Malaysia, Malaysia.

Chen, T. H., Chung, H. Y., Wang, D. H., and Lin, S. H. (2009). Growth and biomass of makino bamboo in Shihmen reservoir watershed area. Quarterly Journal of Chinese Forestry 42:519-527.

Chung, H.Y., Liu, C. P., and Chen, T. H. (2010). Effects of thinning on the growth and biomass of *MAKINO* bamboos in Lienhuachih. Quarterly Journal of Chinese Forestry 43(2):223-231.

Dai, K. Y., Yang, P. L., and San, Z. C. (1973). Bamboo resources in Taiwan. JCCR, TFB Cooperation Report.

Jiang, Z. H. (2002). Bamboo and rattan in the world. Liaoning Science and Technology Publishing House. China.

Kao, Y. P., Lin, W. C., and Chang, T. Y. (1989). The propagation of *Dendrocalamus giganteus* and *D. asper* by level culm-cuttings. Bull. Taiwan For. Res. Inst. New Series 4(2):53-65.

Kiang, T., Lin, W. C., Kang, Z. Y., and Hwang, S. G. (1976). Comparative study on the production of bamboo shoot from the eight strains of *Dendrocalamus latiflorus*. Quarterly Journal of Chinese Forestry 9(1):1-7.

Lin, S.H., Lin, D.G., and Chen, T. H. (2009). The growth characteristics and landslide expansion potential of Makino bamboo forest. Journal of Engineering Environment 23:25-40.

Lin, N. S., Chen, M. J., Kiang, T., and Lin, W. C. (1979). Preliminary studies on bamboo mosaic disease in Taiwan. Bulletin No. 317. Taiwan Forestry Research Institute.

Lin, W. C. (1958). Studies on the growth of bamboo species in Taiwan. Bulletin No. 54. Taiwan Forestry Research Institute. 29pp.

Lin, W. C. (1961). Study on the classification of Bambusaceae in Taiwan. Bulletin No. 69. Taiwan Forestry Research Institute. 145pp

Lin, W. C. (1962). Studies on the propagation by level cuttings of various bamboos (I). Bulletin No. 69. Taiwan Forestry Research Institute. 48pp

Lin, W. C. (1964). Studies on the propagation by level cuttings of various bamboos (II). Bulletin No. 105. Taiwan Forestry Research Institute. 52pp.

Lin, W. C. (1967). Bamboo species and distribution in Taiwan. Journal of Taiwan 3(2):1-20.

Lin, W. C. (1974). Studies on morphology of bamboo flowers. Bulletin No. 248. Taiwan Forestry Research Institute. 117pp.

Lin, W. C. (1976). The classification of subfamily Bambusoideae in Taiwan (continued). Bulletin No. 271. Taiwan Forestry Research Institute. 75pp

Lin, W. C., Kang, Z. Y., Hwang, S. G., and Kiang, T. (1962). Investigation on resources of important bamboos in Taiwan. Cooperative Report no. 4. Taiwan Forestry Research Institute. 28pp.

Lin, W. C., Kiang, T., and Chang, T. Y. (1976). Study on introduction and utilization of *Melocanna baccifera* in Taiwan. Bulletin No. 281. Taiwan Forestry Research Institute. 17pp.

Lin, W. C., Kiang, T., and Chang, T. Y. (1977). Studies on introduction and utilization of giant bamboo(*Dendrocalamus giganteus*). Bulletin No. 300. Taiwan Forestry Research Institute. 18pp.

Liu, S. C., and Kao, Y. P. (1988). Generalized biomass equations for moso bamboo and leucaena plantations. Bull. Taiwan For. Res. Inst.New Series, 3(1):393-406.

Lobovikov, M., Paudei, S., Piazza, M., Ren, H., and Wu, J. (2007). World bamboo resources: A thematic study prepared in the framework of the Global Forest Resources Assessment 2005. FOA, Rome, 2007.

Lu, S. Y., Liu, C. P., Hwang, L. S., and Wang, C. H. (2007). Hydrological characteristics of a Makino bamboo woodland in Central Taiwan. Taiwan J. For. Sci. 22(1):81-93.

Lü, C. M., Liu, C. C., and Rin, U. C. (1982). Experiments on regeneration and improvement of cultivation methods in Moso bamboo stands. Bull 367.Taiwan For. Res Institute.

Lü, C. M., and Chen, T. H. (1992). The structure and biomass of Makino Bamboo (*PhyllostachysMakino)* stand –an example of Tung-Tou area. Bull. Taiwan For. Res. Inst. New Series 7(1):1-13.

Lü, C. M. (1996). The classification of rhizomes in bamboo. Modern Silviculture 12(1):73-90

Lü, C. M. (2000). The build of bamboo forest and reconstruction on damaged area. Modern Silviculture 15(2):52-57

Lü, C. M. (2001). Cultivation and management of bamboo forests. TFRI Extension Series #135. Taiwan Forestry Research Institute. 204pp.

Ueda, K. (1960). Studies on the physiology of bamboo. The Kyoto University Forests. 167pp.

Ueda, K. (1963). A useful bamboo and shoot – a new cultivation technique. Hakusya.

Watanabe, M. (2005). Expansion of bamboo forests in Japan. http://www.kyoto.zaq.ne.jp/dkakd107/frame2.html

Chapter 2

Management and Environmental Conservation of Bamboo Forests on the Slopeland in Taiwan

Shin-Hwei Lin

National Chung-Hsing University

Types of Bamboo Forest and the Distribution on the Slopeland in Taiwan

There are more than 1,200 species of bamboos in the world. Asia has the highest bamboo production area followed by South American, Africa, and Australia. North America and Europe have only short types of bamboo. According to the record on "Flora of Taiwan (2003)", there are fifteen genera, forty species, three varieties, and ten cultivated species of bamboo in Taiwan. According to the statistics of Forestry Bureau, Council of Agriculture, Executive Yuan, the area of bamboo forest in Taiwan was 149,516 hectares in the end of 2000 (not including *Yushania niitakayamensis* and *Sinobambusa kunishii* (Kunishi Cane) in high mountain area), which was 7.11 per cent of the total area of forests (2,101,719 hectares) in Taiwan. The bamboo areas in various counties and cities are listed in Table 2.1.

Among the counties and cities, Nantou County has the largest area of 22,310 hectares, including Jhushan Township, Lugu Township, and Sinyi Township. Gukeng Township in Yunlin County and Meishan Township in Chiayi County produced the most quantity of bamboos. Other countries with high bamboo areas are, 21,713 hectares (14.2 per cent) in Chiayi County, 16,074 hectares (10.7 per cent) in Tainan County, 15,334 hectares (10 per cent) in Miaoli County and Kaohsiung County

15319 hectares, and 9,591 hectares (7.2 per cent) in Hsin Chu County. There are around 17,800 hectares of *Fargesia spathacea* grasslands in mountain areas, mainly *Yushania niitakayamensis* and Kunishi Cane, which form the alpine landscape.

Table 2.1: Area under Bamboo Forests in Counties and Cities in Taiwan.

County or City	Area (Hectare)
Taipei City	565
Kaohsiung City	109
Taipei County	9,548
Yilan County	2,648
Keelung City	717
Taoyuan County	6,519
Hsin Chu County	9,591
Miaoli County	15,334
Taichung County	4,361
Changhua County	1,171
Nantou County	22,310
Taichung City	1,551
Yunlin County	7,033
Chiayi County	21,713
Tainan County	16,074
Chiayi City	87
Kaohsiung County	15,319
Pingtung County	918
Taitung County	5,350
Hualien County	8,598
Total	**149,516**

Source : Forestry Bureau, 2000.

Regarding the distribution of the six economically important bamboo species (Table 2.3), Moso Bamboo (*Phyllostachys pubescens* Mazel *ex* Houz.) are distributed in higher mountains of low and medium elevation around 600-1,500 meters . The other five species are distributed in plain areas.

Bamboo forests in Taiwan are established as small-scale cultivation areas of about 0.5 hectares and below. Makino Bamboo (*Phyllostachys makinoi* Hayata) are mainly cultivated in northern areas like Taipei and Hisn Chu. Other central areas like Nantou are also included. Thorn Bamboo (*Bambusa stenostachya* Hackel) are primary cultivated in Tainan and Kaohsiung. Other subterranean bamboo species, such as Ma Bamboo (*Dendrocalamus latiflorus* Munro), Long-Branch Bamboo (*Bambusa dolichoclada* Hayata), and Green Bamboo (*Bambusa oldhamii* Munro) are also cultivated. *Dendrocalamus latiflorus* is mostly cultivated in central areas.

The Management of Bamboo Forests and the Usage of Bamboo in Taiwan

1. General Management Features of Bamboo Forests in Taiwan

Since the managing period for bamboo forests is short and small areas are required for cultivation, intensive farming is generally adopted. The purposes of managing bamboo forests in Taiwan include producing timber, growing bamboo shoots, and land conservation. Early in 1960-1970, the bamboo management was dependent on Joint Commission on Rural Reconstruction which emphasized on developing and promoting various studies on bamboo species, guiding farmers the technologies of bamboo forest improvement, and establishing the production prefectures of bamboo shoots. Besides, in order to rapidly realize the benefits, the government encouraged farmers to plant bamboo forests. In recent years, the economic value of bamboos is declining and farmers gradually have reduced the operation and management of bamboo forests resulting in mismanagement of the bamboo forests.

Table 2.2: Management of Six Important Bamboo Species in Taiwan.

Bamboo Species	Features	Management purpose
Makino Bamboo (*Phyllostachys makinoi* Hayata)	Leptomorph type rhizomes. Rapid grwoth. It is estimated that Makino Bamboo can grow 2.45 cm a day.	Mainly for bamboo timber, and also for harvesting bamboo shoots.
Moso Bamboo (*Phyllostachys pubescens* Mazel *ex* Houz.)	Leptomorph type rhizomes. Having high cold-resistance and wind-resistance, but not temperature-resistance.	
Ma Bamboo (*Dendrocalamus latiflorus* Munro)	Pachymorph type rhizomes. Most widely cultivated for bamboo shoot in Taiwan.	Mainly for harvesting bamboo shoots, and also for producing bamboo timber.
Green Bamboo (*Bambusa oldhamii* Munro)	Pachymorph type rhizomes. With high heat-resistance, they can grow in high-temperature and damp areas as well as sandy loam and loamy sand soils.	
Thorn Bamboo (*Bambusa stenostachya* Hackel)	Pachymorph type rhizomes. Planted in coastal or dry areas for windbreak or sand stabili-zation; or along streams to protect the revetment.	Mainly for bamboo timber, and focusing on the function of conservation.
Long-Branch Bamboo (*Bambusa dolichoclada* Hayata)	Pachymorph type rhizomes. Planted along stream or around the farmland for environment conservation. Good material for architecture, agricultural tools, equipment and handcrafts.	

In recent years, the usage of bamboos has mostly been replaced by plastics. Therefore, the annual harvest area is only 200 hectares in the entire island of Taiwan. With the reduction of economic production of bamboos, many private bamboo business firms have moved to Southeast Asia or Mainland China where the wages are lower. Furthermore, bamboo forests along riverbanks or in wilderness areas or abandoned bamboo forests are disintegrated into pieces with diesel integrator to become organic fertilizers. Moreover, some disintegrated bamboo timbers are delivered to paper mills as the material of paper pulp. Figure 2.1 presents the Thorn Bamboo being disintegrated into the material of paper pulp.

2. Bamboo for Soil and Water Conservation

Bamboos are suitable for afforestation as they grow fast and their subterranean stems are having network structure, which can stabilize the landslide or seriously erodible areas. The plantation of Thorn Bamboo or Long-Branch Bamboo is undertaken around farmland to check wind erosion, protect the revetment and resist intrusion along the gully and the torrent. Besides, staking and wattling are the primary revegetation and protection methods on slopeland in Taiwan where Thorn Bamboo, Long-Branch Bamboo, and Moso Bamboo, are used.

Figure 2.1: Chips of Thorn Bamboo making the Material for Paper Pulp.

In order to reduce the coastal wind erosion disaster and protect farmland crops in coastal areas, protection through bamboos was provided by creating sand control hedges for disaster mitigation. Since there are abundant bamboos in Taiwan, which are cheap and can be mass produced and harvested, the common bamboo constructions could be made from bamboos pieces, culms or their derivatives.

Figure 2.2: Long-Branch Bamboo along the Gully and the Torrent to Protect the Revetment.

Features of Bamboo Forests

Considering the abundance of Makino Bamboo forests in northern areas and Thorn Bamboo forests in southern areas in Taiwan, the habitats and environmental conservation function of these forests have been described below:

Figure 2.4: Bamboo's Culm and Pieces Used on Staking and Wattling

1. Makino Bamboo (*Phyllostachys makinoi* Hayata) Forests

(1) Characteristics of the Habitats

Makino Bamboo is a native species of Taiwan, which grows in the plain areas at sea level as well as in mountain areas upto 1,500 meters above sea level. Makino Bamboo was cultivated in the early years, and now it is distributed widely in northern Taiwan. Temperature is the key factor for bamboo buds and shoot growth. Rhizome rapidly extends especially in the bared soil or landslide areas. Makino Bamboo originally belongs to temperate zone. Therefore, high temperature is adverse for its growth. Makino Bamboo prefers steady climate, thick soil, abundant nutrient, sufficient annual rainfall, and good drainage. Because of its tolerance to the environmental stress is strong it can grow in the places which are cold, hot as well as barren and dry land. However, in these stress habitats, it does not make a good forest.

Compared to the common tree forests, Makino Bamboo forests are with less diversity. Woody plants

Figure 2.4: Bamboo Shoot (Chip) and Fascine Used for Sand Stabilization in Coastal Areas.

Figure 2.5: Bamboo Shoot (Chip) Used for Wind-Protection in Coastal Areas.

Figure 2.6: Makino Bamboo Forest in the Catchment of Shihman Reservoir.

growing with Makino Bamboo forests include *Callicarpa formosana, Arenga tremula, Tetrapanax papyriferus* and *Musa formosana;* and herbaceous vegetation contains *Cyclosorus parasitica, Pteris ensiformis, Alocasia Macrorrhiza, Miscanthus floridus* and *Arundo formosana.* The leaves of Makino Bamboo are the food for nympha of *Nymphalidae,* such as *Lethe chandica Moore, Lethe europa, Neope muirheadi nagasawae Matsumura, Penthema formosanum,* and *Stichophthalma howqua formosana Fruhstorfer* as well as *Hesperiidae,* like *Telicota bambusae horisha* and *Telicota ohara formosana.* The amphibian and reptile fauna in Makino Bamboo forests are less, such as Scincidae and Rhacophoridae which often hide and breed in Makino Bamboo forests.

(2) Physiological and Growth Characteristics of the Root System

Makino bamboo has extremely high productivity due to its laterally widespread root system. The root system has abundant nutrition storage and supplies nutrients adequately for the quick growth of the bamboo stem. Makino bamboo has a running rhizome system and individual stems emerge from the rhizomes at different intervals. In addition, rhizomes constantly show a very high growth rate and this ensures a rapid vegetative reproduction of this species.

The stem culm and root system of Makino bamboo have a life cycle of about 10 years and the Makino bamboo shows very high germination following seeding. The growth rate is very high when the culm age between 3 and 5 years. Due to high

Figure 2.7: Diagram of Ecological Components in Habitats of Makino Bamboo Forests.

density of the root and high growth rate of the stem, the invasion and succession of broadleaf trees into the bamboo forest are almost impossible. In particular, during the first invasion into barren land, the bamboo exhibits an exceedingly high expansion potential to prohibit other plant species to colonize. This may cause some forest gaps devoid of trees especially for bamboo forests older than 8 years. In addition, defoliation of Makino bamboo during the autumn and winter gives an impression of the fruit as an open fruit.

The fibrous roots are dense and decrease with the depth of soil. From the field survey, the fibrous roots of Makino Bamboo, with the diameter around 0.1-0.4cm, were found to be distributed densely in the soil depth of 0-30cm, followed by 30-60cm. The rhizome, with the diameter around 1-3 cm were, mostly distributed in the soil depth 0-30cm, and only part of them would come out of soil layer. Since the rhizomes grow in horizontal directions, they criss-cross each other and grow like a net, and entangle together in the soil depth of 10-20cm. The infiltration rate of the shallow rooted layer is high and it is very low in lower non-rooted layer or not permeable. This results in the enhanced sub-surface flow of water along the slope. Thus, bamboo forests in catchments influence the watershed positively through water conservation, water purification, and sediment prevention in reservoirs.

The characteristics of Makino Bamboo root system and the relevant parameters are shown in Table 2.4.

(3) Expansion Prevention of Makino Bamboo Forests

Expansion prevention and renewal forestation of Makino Bamboo forests in the catchments of reservoirs are difficult. Selected cutting along the contour and eliminating new bamboos are recommend and can effectively restrain the expansion of Makino Bamboo forests. However, it could only be used in the edge of Bamboo forests or in the scattered plots. For the management of aged and degraded Makino Bamboo forests, a healthy forest in barren areas around the bamboo forests may be created to reduce the expansion effect of bamboo forests. Besides, suitable slopeland protection engineering or drainage construction or the landslide treatment methods are available to prevent the expansion of bamboo forests in catchments.

(4) Harvest and Conservation of Bamboo Forest

During the cutting process of Makino Bamboo, the problems of soil erosion and slope stabilization should be taken into account. The peak period of cutting is at the end of bamboo shoot production or before New Year. Nevertheless, the end of bamboo shoots production (May-June) is followed with typhoon season (July-August) when the bare land of Makino Bamboo forests might be exposed to high erodibility. Furthermore, most cuttings of Makino Bamboo are focused on reachable side of the slope near the roads. In Makino Bamboo forests, density management and cutting of bamboos of different ages are less. Therefore, availability of main bamboos is few. Since the price of Makino Bamboo has dropped in recent years, the cutting has become rough. Only the bamboos along the slopes are cut and the cut timbers are allowed to slip along the slope to reach the road in order to reduce the trucking cost. This causes land erosion along the barren slope.

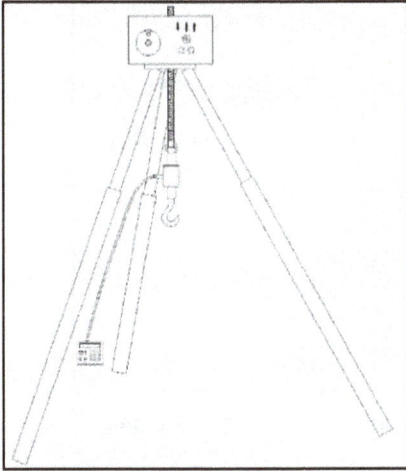

Layout of pull-out equipment

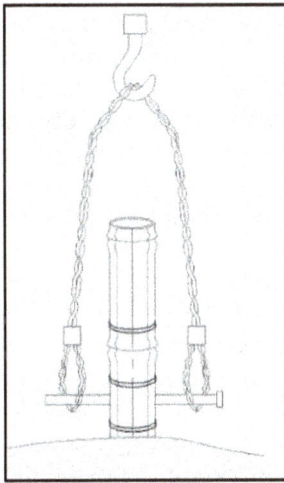

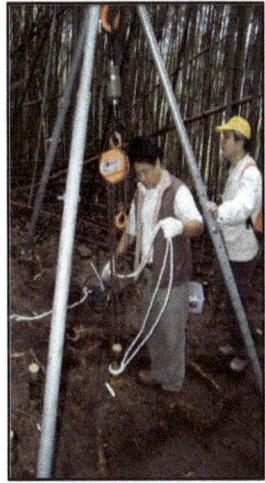

Connection of root system with pull-out system

Figure 2.8: Pull-out Equipment and its Connection with Makino Bamboo Root System.

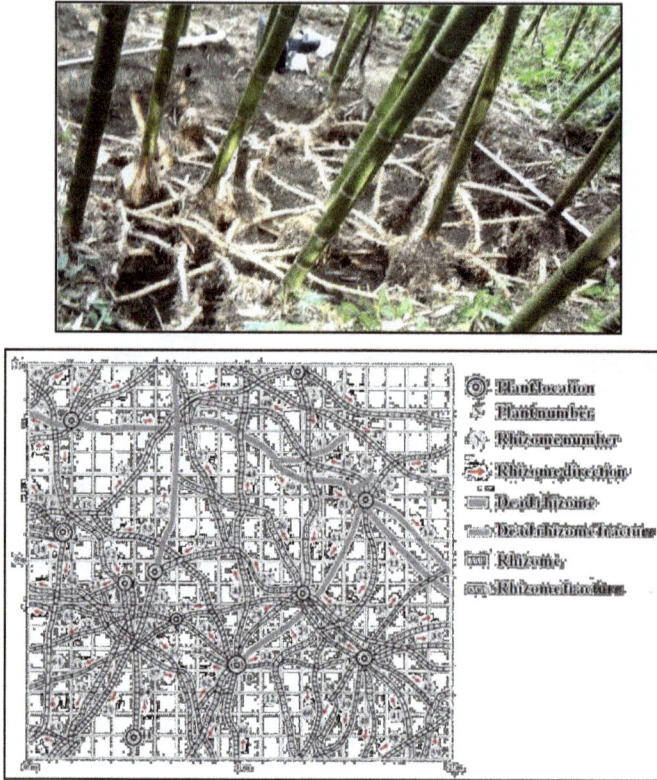

Figure 2.9: Distribution of Makino Bamboo Root System.

(5) Typical Landslide Process in the Makino Bamboo Forest

The ground part of Makino Bamboo forest is tough culms, while the underground part is network root system. The rapid growth of subterranean rhizomes forming network root system provides anchorage to bamboo plants. With fast growing fibrous dense root system on top soil layer, the rhizome (main horizontal root system) dominates 0-15cm soil depth and the large amount or fibrous roots could reach the lower soil up to 60-70cm depth. To sum up, according to the observation on soil strengthening by Makino Bamboo fibrous roots, it is found that the soil reinforcement function of Makino Bamboo in shallow layer is great, but cannot parallel with that of general woody plants in the lower layer. Since, the root systems of Makino Bamboo forests are shallower than other woody plants, landslides are likely to occur especially in steep slopeland or in riparian zone. The shallow-rooted bamboo forests can result seepage on the interface between rooted and non-rooted systems, and soil layer can become free and that induces headward erosion. When the drainage problem of free end is not well handled and the slope is exposed to rainfall erosion, the basal slope will be destroyed, and damages from landslides will be aggravat ed further.

Table 2.4: Features of Makino Bamboo's Root System and the Relevant Parameters.

Item	Bamboo sps.	Makino Bamboo Forests
On the ground	Height	Culm 10.8±1.8 meters
(stem leaves)	Density	8,000~12,000 culms/hectare
	Litter	4.33 tons/hectare/year
	Weight on the ground	157.2±31.2 tons/hectare
	(Rainfall) Interception storage	Low
	Ground cover plants	Few
Under the ground (root system)	Density	Shallow layer, (0-60cm) density about 0.09~0.44 per cent (not included root diameter below 1mm)
	Type	Long creeping rhizomes bearing monopodial culms
	Depth	Rhizomes: 0~40cmFibrous roots: 0~60cm with up to 70cm depth
	Weight	291.54 tons/hectare(wet weight)
	T/R ratio	1/1~2/1
Root field soil	Infiltration	The capacity of root layer soil is high, while it is lower in the non-rooted layer. Abscission layer often appears on soil-root interface.
	Profile	The development of soil profile is not completed.
	Soil moisture	Higher in summer, very low in winter (dry season)
	Surficial erosion	Less
Habitat	Sustainability	Small (short)
	biodiversity	Low
	Landscape harmony	Low harmony, or merely some parts with landscaping function
Landslide	Landslide rate	No significant difference (competition with other of forest type in the large-scale catchment), the effects of natural force and man-made interruption are larger influence.
	Types	Shallow landslide, in chips or small pieces
	Follow-up landslide at landslide periphery	Headwater erosion, soil-root abscission layer, small-scale landslide periodically.
Usage and	Economic effects	Low
management	Environmental function	Low

Source. Lu, Chin-Ming (2001), Lin, Hsin-hui (2001), Chen, Yao-jung (2006), Chen, Tsai-hui (2007), and Lu, Hsiang-yu (1988, 2003, 2007).

The landslides of bamboo forests on slopeland are resulted from the gap caused by runoff erosion on the surface soil layer. Presently, most landslides of bamboo forests are shallow or slip landslides with the features as follows.

Figure 2.10: Landslide Treatment Areas around the Bamboo Forest.

Figure 2.11: Clear Cutting (Timbering) of Makino Bamboo Forests.

**Figure 2.12: Clear Cutting (Timbering) and following Rapid
Regeneration of Makino Bamboo Forest.**

Figure 2.13: Diagram of follow up Landslide Process in Makino Bamboo Forest.

1. Since the basal slope is destroyed, the rainfall erosion has resulted in slope cut bareness.

2. The root systems of Makino Bamboo are dense and shallow. The basement soil are likely to be washed away causing the root systems of Makino Bamboo to hang in the air.

Contd...

Figure 2.13–*Contd...*

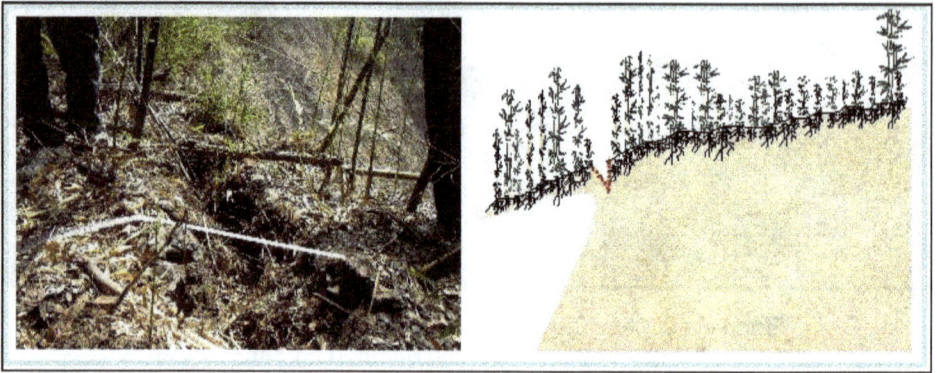

3. The tension gap appears on the slopeland surface of Makino Bamboo forest from landslide or weight load. The rains accelerate the erosion.

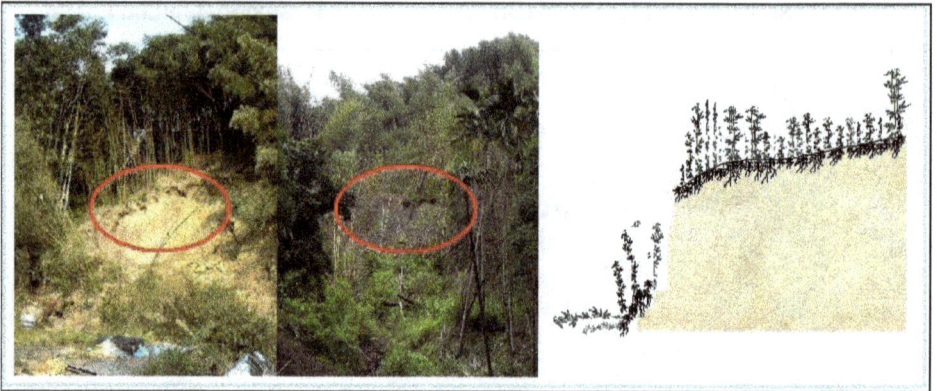

4. Finally, it cannot bear the weight load so that it starts to slide. If it is not controlled, the landslides will occur periodically.

2. Thorn Bamboo (*Bambusa stenostachya* Hackel) Forests

(1) Characteristics of Thorn Bamboo Forests (Especially on the Mudstone Areas)

The tolerance of Thorn Bamboo is quite strong. It can grow in dry, high saline, and nutrient poor soil. Mudstone areas in south-western Taiwan have roads in mountain areas constructed in the past forty years. Besides, with disturbance and destruction of mountains as well as overuse and irrational management of lands, the soil erosion of mudstone areas has been accelerated. The bare mudstone erodes rapidly increasing the barren areas where plants are not easy to grow. For fast reforestation in early stage, the government encouraged farmers to plant Thorn Bamboo, which would not change the growing locations. Once it grows, new bamboo shoots would

Figure 2.14: Plants do not easily grow in Mudstone Area.

Figure 2.15: Thorn Bamboo could grow in the Gully Alluvial Soil of Mudstone Area.

Figure 2.16: Thorn Bamboo and Long-Branch Bamboo Forest in Wusanto Reservoir Catchment.

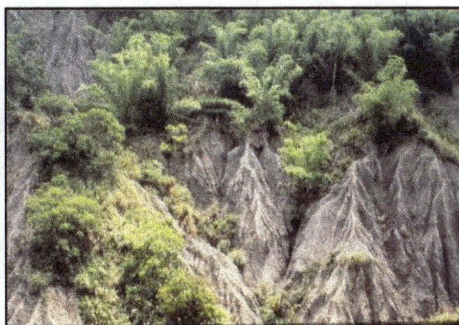

Figure 2.17: Gully Erosion in Thorn Bamboo Forest.

continuously be sprouting out of rhizome. For this reason, Thorn Bamboo has become the pure stand forest in mudstone areas in south-western Taiwan with the plantation area being 80 per cent of the reforestation area.

(2) Features of Root System of Thorn Bamboo

Thorn Bamboo root system does not show large variation among the clumps. The diameter of vertical root cylinder at the base

Figure 2.18: Large Spacing between Thorn Bamboo Clump.

is about 2-3 meters and the spacing is about 10-15m between two adjacent bamboo clumps. Because the root clump spacing is too large, the surface soil in the non-rooted area could not be held by buttressing of the root cylinder, and the arching effect is quite low especially on the mudstone areas. Therefore, serious soil erosion by surface runoff happens in the slopeland of Thorn Bamboo forest.

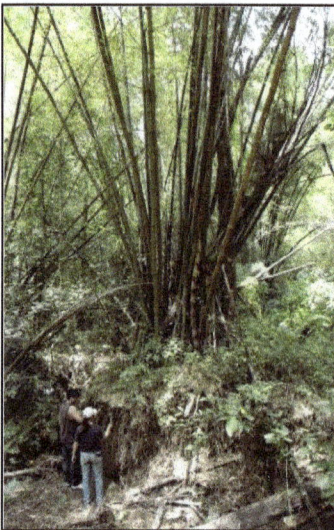

Figure 2.19: Thorn Bamboo and It's Root System.

Figure 2.20: Root System of Thorn Bamboo (Close Shot).

The fibrous root can reach 3 meters depth, but mostly it is distributed within 100cm depth. The sub-surface layer within 10cm depth has high density fibrous root system while the average diameter is about 0.1cm.

(3) Fire Problem in Thorn Bamboo Forests

Dry branches and dead leaves of Thorn Bamboo often cover the ground of the forests. The friction caused by wind and man-made spark could easily result in fire in

Figure 2.21: Litter on the Floor of Thorn Bamboo Forest can Induce Forest Fire.

bamboo forests. Fire affects the composition and structure of forest vegetation causing disturbance in the ecological system. Once the forest vegetation is affected by fire, it is hard to control. The allelopathy of Thorn Bamboo is quite high. Therefore, a few species can invade the forest. Fire on Thorn Bamboo forest easily spreads resulting in large-scale fire disaster. The fire can continue for 4-5 days, once the fire occurs. Moreover, after the fire, canopy layer will not restore rapidly that intercepts rainfall and reduces gully erosion in the forest.

Figure 2.22: New Sprouting Bamboos and Dead Bamboos Alternatively Grow after Fire.

(4) Typical Landslide Process in the Slopeland of Thorn Bamboo Forest

With the functions of covering and protecting, cohering and solidifying soil body, as well as soil genesis, Thorn Bamboo is an excellent pioneer plant for preventing the erosion and sediment control of mudstone areas. However, with the dense litter, canopy and root system of Thorn Bamboo, other plants can hardly grow in the forests. With the lack of canopy cover water can easily run off so that the functions of water conservation, hydrological regulation, and flood detention are far less than common forests.

The fibrous roots of Thorn Bamboo do not bind soil effectively and the intervals between Thorn Bamboo clumps are large. The soil in the intervals is easily eroded by rainfall to induce gully erosion. Rainfall can cause Thorn Bamboo slopes uncovered and would wash away the soil under Thorn Bamboo so that the root system is exposed. With the high culms and shallow roots, Thorn Bamboo are easily shook by wind that would result in slope slides or extension of slides with the features as shown in Figure 2.23.

1. Thorn Bamboo was planted in the early stage because of its highly adaptation on mudstone areas

2. Cut slope on roadside or river bank erosion areas can induce sliding of Thorn Bamboo clump.

3. Thorn Bamboo clumps slide down to the lower end of the slope because of the surface erosion and the root cylinder is bared.

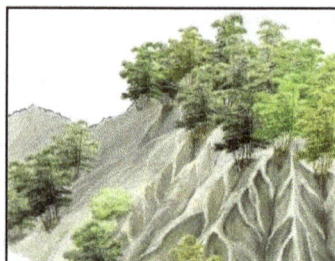

4. Gully and headward erosion that cause the bamboo stumps slide periodically.

5. The slopelands remain largely in bared condition on the steeper surface, but the slided bamboo stump could continuously grow on the lower end of the slopes.

Figure 2.23: Landslide Process in the Mudstone Slopeland of Thorn Bamboo Forest.

References

Biodiversity Center, National Taiwan University (2006). Database of Natural Sources and Ecology in Taiwan–III Agriculture, Forestry, Fishery and Animal Husbandry. Taiwan Forestry Research Institute, Council of Agriculture, Executive Yuan.

Chang, C. P. (2004). Research on the landscape ecological structure and change in mudstone areas. Department of Soil and Water Conservation, National Chung Hsing University, PhD dissertation.

Chen, Y. J. (2006). Research on *Phyllostachys makinoi* landslide mechanism. Department of Soil and Water Conservation, National Chung Hsing University, Master's dissertation.

Flora of Taiwan Second Edition Board of editors (2003). Flora of Taiwan second edition (sixth volume). College of Life Science, National Taiwan University.

Hsieh, M. T. (2009). Research on root system of *phyllostachys makinoi* and features of the roots. Department of Soil and Water Conservation, National Chung Hsing University, Master's dissertation.

Kao, C. C. (1998). Research on drought tolerance of *Thorn Bamboo* and the roots in south-western mudstone areas, Department of Soil and Water Conservation, National Chung Hsing University, Master's dissertation.

Liang, A (1982). Bamboos. Taipei: Harvest.

Lin, H. H. (2006). Reports on the "Establishment of Mudstone Areas Conservation and Community Landscape Planning". Soil and Water Conservation Bureau, Council of Agriculture, Executive Yuan.

Lin, H. H. (2008). Application Manual for Catchment Vegetation. Soil and Water Conservation Bureau, Council of Agriculture, Executive Yuan.

Lin, H. H. (2009). Discussion of *Phyllostachys makinoi* forest features and landslide mechanism. 2009 Shihmen Reservoir and the Catchment Remediation Project Result Exchange Seminar, Taoyuan County.

Lin, W. C. (1996). Bamboo Species Collections of Lin, Wei-chih. Taiwan Forestry Research Institute, Council of Agriculture, Executive Yuan, Taipei City.

Lin, W. C., Kang, T. J., Huang, S. K., and Chiang, T. (1962). Investigation on primary bamboo sources in Taiwan. *Taiwan Forestry Research Institute Corporate Report No. 4.*

Liu, P. H. (1987). Cultivation and usage of bamboos. Taipei City: Wu Chou Press.

Lu, C. M. (2001). Cultivation and Management of Bamboo Forests. Taipei City: Taiwan Forestry Research Institute, Council of Agriculture, Executive Yuan.

National Chung Hsing University (2006). Reports on catchments in national compartment regulated with ecological engineering. Taiwan Forestry Research Institute, Council of Agriculture, Executive Yuan.

Tai, K. Y., Yang, P, L., and Shen, J. C. (1973). Taiwan bamboo forest resources. Corporate Project of Joint Commission on Rural Reconstruction, Forestry Bureau, Aerial Survey Team, National Pingtung Agriculture College pp.82.

Taiwan Endemic Species Research Institute, Council of Agriculture (2010). Hushan Reservoir Engineering Project Ecology Conservation Measures – Final Report of Investigation and Research on Ecological System of Forests and Rivers. Central Region Water Resources Office, Water Resource Agency, Ministry of Economic Affairs.

Tseng, L. L. Life plants – Usage of bamboos. http://web.ptes.tp.edu.tw/

Ueda, K. (1963). Bamboo Shoots of Utilized Bamboos – New Technology of Cultivation. PHP.

Water Reources Agency, MOEA (2009). Reports on Management of Habitat in Reservoir Catchment and Regulation and Conservation Strategies (2/2). Water Resources Agency, MOEA.

Water Resources Agency, MOEA (2009). Reference manual for management of habitat in reservoir catchment and regulation and conservation strategies. Water Resources Agency, MOEA.

Water Resources Planning Institute, Water Resources Agency, MOEA (2009). Reference manual for seashore vegetation plan. Water Resources Planning Institute, Water Resources Agency, MOEA.

Yen, T. M. (2004). Management of bamboo forests handouts. Taichung City: Department of Forestry, National Chung Hsing University.

Chapter 3

History, Current Status, and Prospects of the Bamboo Industry in Taiwan

Yu-Jen Lin

Division of Forest Utilization, Taiwan Forestry Research Institute, 53 Nanhai Rd., Taipei 10066, Taiwan

Introduction

There are abundant bamboo resources accounting for 7.2 per cent of the total forest area in Taiwan. The distribution of bamboo in Taiwan can be both natural and planted, and ranges from the flatlands to high-mountain areas (0-3000 m) with 85 species within 15 genera (Lü, 2001). There are 6 main commercial bamboo species in Taiwan: Moso bamboo (*Phyllostachys pubescens*), Makino bamboo (*P. makinoi*), ma bamboo (*Dendrocalamus latiflorus*), thorny bamboo (*Bambusa stenostachya*), long-branch bamboo (*B. dolichoclada* Hayata), and green bamboo (*B. oldhamii*) (Lü, 2001). Each species plays an important role in bamboo utilization and offers local people a variety of products for their daily needs and helps protect habitats.

Bamboo utilization has a long and turbulent history in Taiwan. In the 1960~1980s, bamboo-related industries significantly contributed to local economies, providing jobs and revenue to bamboo farmers, local communities, and the government. But the industry declined due to soaring labor costs and high competition by cheaper imported products. Consequently, most processing factories shifted to China and Southeast Asia since the 1980s to reduce production costs. For decades now, the economic potential of bamboo has been largely neglected.

However, in recent years, bamboo has enjoyed renewed attention because of its rapid growth and high productivity. These make it an interesting option for biomass, especially as climate change concerns and energy costs soar (INBAR 2009, Windenoja 2007). Currently, strategies for reviving the bamboo-processing industry and bamboo production so that bamboo is better utilized, and to maintain the health of indigenous bamboo forest resources are important issues for Taiwanese agricultural authorities.

Through collecting relevant statistics and reviewing published articles that analyzed bamboo utilization and the bamboo-processing industry in Taiwan, this paper first illustrates the historical development of the bamboo-processing industry in Taiwan, then describes the current status of this declined industry, and finally mentions several beneficial programs by related governmental agencies to upgrade the bamboo industry in Taiwan. It also provides the potential development of bamboo utilization with the efforts of the Taiwanese agricultural authorities.

Development History of the Bamboo Industry in Taiwan

Because bamboo is easy to process by hand, since earlier times, bamboo utilization in Taiwan was commonly distributed in farm villages located in areas with abundant bamboo resources. Most bamboo products in earlier times were handmade using simple hand tools. All parts of the entire bamboo plant, including the roots, culm, branches, twigs, leaves, and shoots, can be utilized by humans (Liese 1987, Scurlock *et al.*, 2000). Table 3.1 shows bamboo utilization and related products based on the parts the plant. The greatest utilization is using the culm for agricultural and fishing tools, furniture, construction bridges, scaffolding, building house frames, walls, window frames, roofs, interior dividers, etc. Except for green bamboo, the other 5 main commercial bamboo species in Taiwan can provide the abovementioned functions.

Edible bamboo shoots among the 6 main commercial species are Moso, Makino, ma, and green bamboo. In addition, ma bamboo shoots are processed in a dried form for storage, while the 3 other kinds of shoots are cooked and eaten fresh. The market price of Moso bamboo shoots is highest due to its limited production time, hard harvesting, and unique delicious taste (Table 3.1).

The bamboo-processing industry was highly developed due to the progress of processing technology using machinery, especially the development of lamination technology. The historical development stages of the bamboo-processing industry in Taiwan based on the product of life cycle are described as follows (Lee *et al.*, 1993).

Introduction Stage: Before 1969

Before 1960, bamboo was utilized to make various products for household goods and agriculture appliances, but the quantity of most products did not reach an economic scale because things were handmade. Since the 1960s, the bamboo-processing industries in Taiwan gradually developed because the Japanese bamboo industry shifted to Taiwan, because Japan was facing the problem of high production costs domestically. After that, Taiwanese bamboo products began to be exported abroad employing the advantages of abundant bamboo resources, low wages, and low production costs. At this stage in 1961~1969, the export value of Taiwanese bamboo products increased from US$ 523,000 to 2,453,000.

Table 3.1: Bamboo Utilization and Related Products Sorted by Parts of the Bamboo.

Part of the Bamboo	Related Products
Branches and leaves	Brooms, pan brushes, bamboo hats, leaves for bamboo dumplings, etc.
Culm	Culm: materials for households, materials for building of gardens, scaffolding poles for buildings, furniture manufacturing, supplementary implements for agriculture and fishery.
	Strip making: joss sticks, knitting sticks, bamboo mats, birdcages, light ornaments, skewers, sushi roller mats, toothpicks, bamboo curtains, pan brushes, etc.
	Tools: rice spoons, bamboo swords, ear scratchers, bread clips, combs, chopsticks, etc.
	Sliver intertwines: tea plates, bamboo screens, bamboo ceilings, bamboo wall boards, furniture, vases, handbags, etc.
	Cylinders: pen tubes, deposit tubes, incense tubes, rice tubes, etc.
	Boards made from rotated cut bamboo veneer: coffee plates, tea plates, bamboo screens, bamboo composite floors, bamboo ceilings, furniture, etc.
	Cylinder furniture: chairs, tables, magazine shelves, flower stands, etc.
	Culm bound with string: screens, poles, fences, etc.
	Square slices: seat cushions, car cushions, etc.
	Handmade appliances: vases, handbags, baskets, etc.
	Artwork: using engraving skills to create ornaments and functional products.
Roots	Handles of handbags and other various ornaments.
Shoots	Food.

Source: Lin (2006).

Growth Stage: 1970~1975

In this stage, the export value consistently increased due to contributions of the shift of Japanese processing technologies, many innovations in domestic processing machinery, and several assistance in construction projects that were completed with governmental policy support, such as the bamboo preservation factories, specific parks for the bamboo-processing industry, and bamboo utilization becoming an integrated field. The export value increased from US$ 4.3 million in 1970 to 18.1 million in 1975.

Mature Stage: 1976~1980

This stage was the golden time for the bamboo-processing industry in Taiwan. Besides the installation of various processing factories by the government, the government simultaneously provided financial support and capital credit to help purchase machinery and solve turnover problems of capital at this stage. In addition, in order to obtain more-skillful labor and promote processing technologies, the government carried out training programs though cooperation between educational institutions and processing factories. Under these advantageous circumstances, the export value reached US$ 30.7 million, and the growth rate reached 70 per cent in 1976 compared to 1975. The export value even promptly increased to US$ 44.7 million in 1978 (Figure 3.1). However, in the last years of this stage, the Taiwanese bamboo-processing industry gradually lost its advantages owing to soaring labor costs and was replaced by China and Southeast Asian countries.

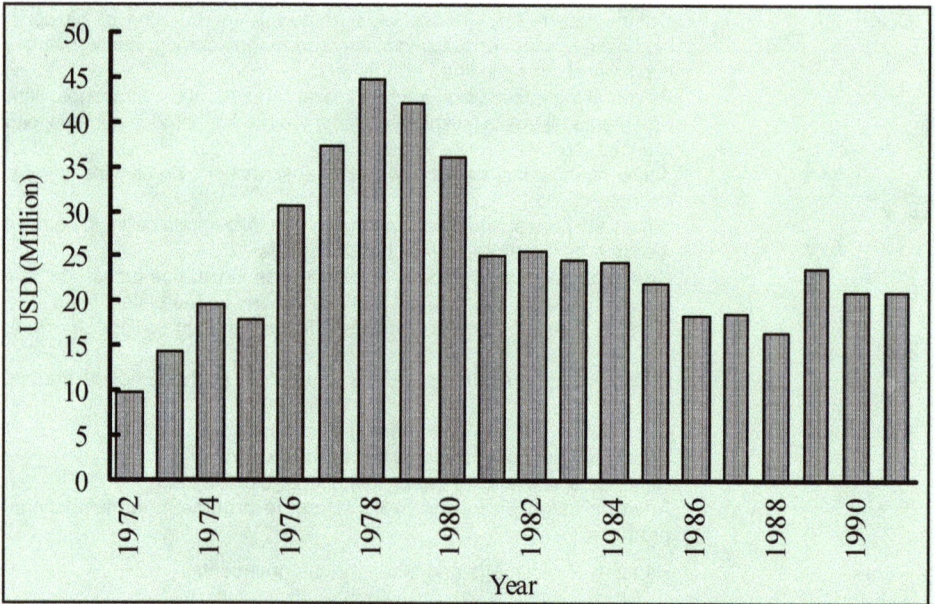

Figure 3.1: Statistics of export value of bamboo products during 1972~1991 in Taiwan.

Declining Stage: after 1982

The golden time of the Taiwanese bamboo-processing industry lasted only about 10 yrs and promptly declined after 1982 not only because of soaring labor costs, the lack of labor and high production costs, but also due to impacts from the mass import of semi-finished and whole products, and the exodus of other related industries. In 1991, in contrast to the time of 10 yr earlier, the export value had decreased to only US$ 21.0 million, which was only around 50 per cent of the previous level, and the import value increased to US$ 2.9 million, a growth of about 91 per cent.

Table 3.2 describes decade-wise main bamboo products in Taiwan. The kinds of products manufactured were usually based on the processing technologies developed at that time. The functions of the products before the 1960s were almost all handmade for household goods and agriculture appliances of low value. Products after the 1980s, even as the industry was declining, consisted of many laminated products, engraved articles, and fine artwork, simultaneously produced in great volume by using a lot of advanced machineries and progressive lamination technologies.

Table 3.2: Time Scale of Bamboo Product Development in Taiwan.

Time Period	Main Bamboo Products
Before the 1950s	Building construction: bamboo booths, scaffolds, fences, etc. Agriculture tools: bamboo hats, dustpans, sieve plates, scoop wheels, etc. Fishery tools: rafts, fish baskets, fishing rods, etc. Food appliances: food steamers, chopsticks, rice spoons, etc. Wedding articles: palanquins, baskets for gifts, baskets for wares of needlework, etc. Articles for livelihood: tea plates, flower stands, baskets for vegetable transport, etc.
1950~1960s	Baskets from bamboo twigs, articles of bamboo weaving, etc.
1960~1970s	Bamboo curtains, skewers, bamboo plates, birdcages, teacup pads, jinrickshas, bamboo windbells, etc.
1970~1980s	Bamboo swords, bamboo furniture, coffee plates, bamboo curtains, etc.
1980~1990s	Bamboo lanterns, bamboo swords, laminated products, furniture with engraving, etc.
After the 1990s	Laminated products, engraving articles, fine artwork, etc.

Source: Lin (2004).

Figure 3.2 shows Taiwan's main export countries of bamboo products during 1971~1991. The greatest one among the main export countries in 1976 was the US with a value of around US$ 14,8 million, followed by Japan with a value of around US$ 11,2 million. Since 1981, Japan became the largest importer instead of the US and kept that position until 1990. In 1990, the export value of bamboo products to Korea approached the value of US$ 11,5 million, which for the first time exceeded that to Japan with a value of US$ 11,4 million. Since then, exports of bamboo products gradually increased to Korea and remained at an important position with Japan.

Current Status of the Bamboo Industry in Taiwan

There were around 1200~1500 registered factories in Taiwan doing bamboo processing and related business during the mature stage (Lee *et al.,* 1993). However, since 1982, the number of bamboo-processing factories rapidly declined year by year due to the loss of the advantage of lower labor costs. In 1993, the number of remaining factories was not even one-third compared the mature stage. In 2004, there were fewer than 100 registered factories still in operation as small businesses in Taiwan. Most factories had shifted to China and Southeast Asia to seek cheaper labor and production costs.

Therefore, bamboo harvesting slowly decreased due to reduction in the output value of the bamboo-processing industry, and amounts of relevant studies of the bamboo-processing industry and the statistical work of bamboo production were dropping off; even the official statistical data of bamboo harvesting by the Taiwan Forestry Bureau are hard to obtain since 2003. Figure 3.3 showed a continuous decreasing trend of the production value of bamboo (only culm,) from 1993 to 2002, at only around US$ 137,000 and 200,000 in 2001 and 2002, respectively.

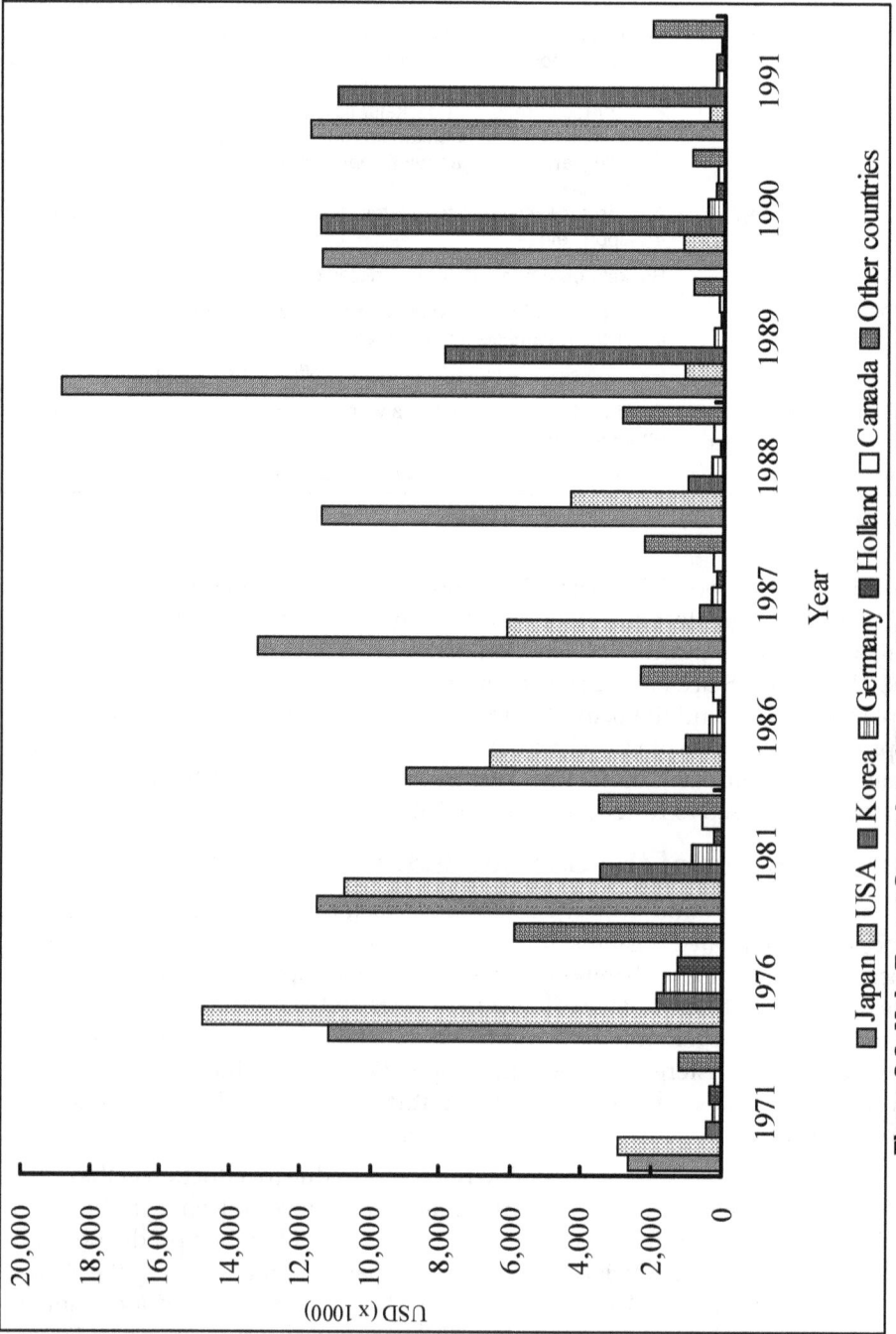

Figure 3.2: Main Export Countries of Bamboo Products during 1971-1991 from Taiwan.

Legend: Japan Korea Germany Holland Canada Other countries

USA

Year axis: 1971 1976 1981 1986 1987 1988 1989 1990 1991

Y-axis: USD (x 1000) — 0, 2,000, 4,000, 6,000, 8,000, 10,000, 12,000, 14,000, 16,000, 18,000, 20,000

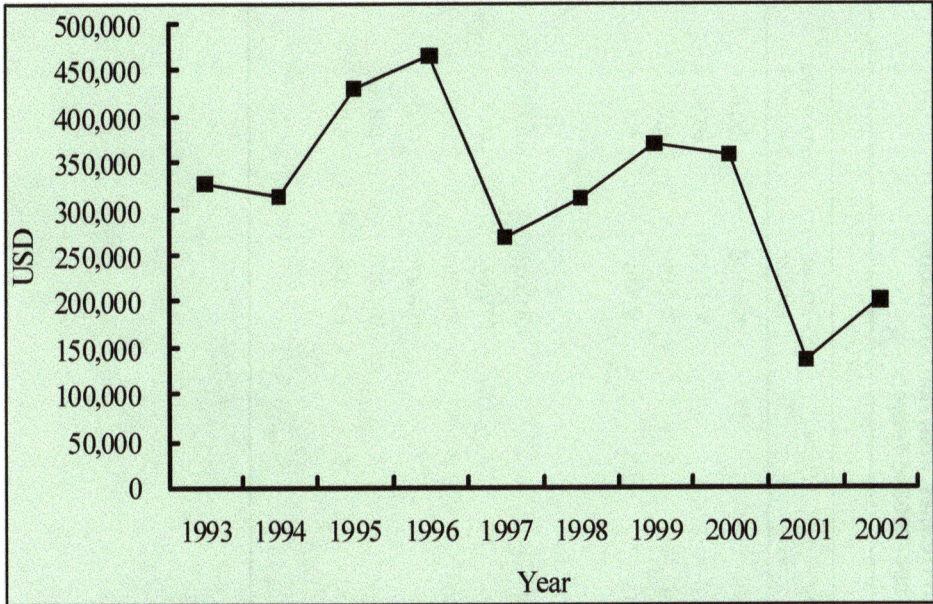

Figure 3.3: Production Value of Bamboo during 1993~2002 in Taiwan.

Because it is in a different category, bamboo shoot production is not included in the bamboo-processing industry in Taiwan. Bamboo shoot production is one of the items among bamboo products for which the production is stable regardless of whether for domestic or foreign markets. The total export value of frozen bamboo shoots overseas in the period of 1994~2003 was worth US$ 21 million, an average of US$ 2.1 million annually, and the export value was around US$ 180,000 in 2003. The main top 3 exporting countries for this product were Japan, the Netherlands, and the USA (Lin 2004).

In addition, despite the bamboo-processing industry having declined in Taiwan, there are still certain demands for bamboo products in the domestic markets for agricultural use and household goods. Amounts of imported bamboo products from China and Southeast Asian countries, such as Vietnam, Thailand and Indonesia, with cheaper prices are substituted for the domestic supply to fill market demand. Table 3.3 is the ranking of countries and output values of bamboo (canes, split skin, and roots) sorted by imports and exports during 1994~2003 in Taiwan. China accounted for 75.3 per cent of the import value during 1994~2002 and reached 98.9 per cent of the import value in 2003 (Lin 2004).

Undoubtedly, China has become the most important import country of bamboo products in Taiwan. With regard to the overall export value being higher than the import value, the reason is that the main import products are raw materials with lower value such as bamboo culm, branches, leaves, and roots, and most export products are processed products with higher value.

Table 3.3: Ranking of Countries and Output Value of Bamboo (Canes, split skin, and roots) Sorted by the Import and Export Value during 1994–2003 in Taiwan.

	Country	Rank	Value (USD)	Percentage (per cent)	Country	Rank	Value (USD)	Percentage (per cent)
			1994–2002				2003	
Import	Overall		15,339,525	100.0	Overall		394,106	100.0
	China	1	11,553,842	75.3	China	1	389,751	98.9
	Vietnam	2	1,428,041	9.3	Vietnam	2	2,245	0.6
	Thailand	3	1,335,379	8.7	Japan	3	2,110	0.5
	Japan	4	527,802	3.4				
	Indonesia	5	354,714	2.3				
Export	Overall		19,663,838	100.0	Overall		1,413,035	100.0
	Japan	1	9,554,882	48.6	China	1	575,625	40.7
	Hong Kong	2	6,575,035	33.4	Hong Kong	2	404,551	28.6
	USA	3	1,233,096	6.3	Japan	3	198,270	14.0
	China	4	668,594	3.4	USA	4	124,406	8.8
	Germany	5	222,795	1.1	Mexico	5	85,206	6.0
	UK	6	203,986	1.0				
	Mexico	7	200,902	1.0				

Source. Lin (2004).

Bamboo grows fast, reproduces prolifically, and matures quickly within 4~5 yr, but the site productivity will rapidly degrade if the bamboo forest lacks suitable forest management for mature culm cutting. Unfortunately, nowadays large areas of unmanaged bamboo forests have formed due to the lack of bamboo cutting for the long term which is a serious waste of natural resources. Therefore, it is critical for the Taiwanese agricultural authorities to promote the amount of bamboo utilization and to restore the bamboo-processing industry and bamboo production once again, so that it can revive the economic activities in villages and recover the prosperity in farms and communities.

To achieve the abovementioned goals, the Taiwanese agricultural authorities have created a project called "Transforming and Reviving Plan of the Bamboo Industry" since 2002. Developing the local bamboo charcoal industry in abundant bamboo forest areas was a major part of the project. Adopting imported technology from Japan, many earthen kilns (a kiln structure is illustrated in Figure 3.4.) for bamboo charcoal production have been built in many locations. The authorities simultaneously assigned experts, who are good at the production technology of bamboo charcoal, to teach bamboo farmers to produce the high-temperature carbonized bamboo charcoal for multi-functional usages.

Bamboo charcoal is made from matured bamboo of over 4 yrs old, which is free from preservatives, termite-proof chemicals, glue, coatings, or other biochemical treatments. High-temperature carbonized bamboo charcoal has powerful absorption/ attachment capabilities due to its dense structure with many pores. Therefore, it can be used to regulate humidity levels and deodorize. Additionally, the specific property of far-infrared radiation emission, which is easily absorbed by the human body, is used to produce textile yarns as well to improve blood circulation and general health (CAS 2006). So, the main function of high-temperature carbonized bamboo charcoal is not to be burned for energy, which is only a kind of utilization of low value.

Using earthen kilns to produce bamboo charcoal requires only a little capital and easy operating technology, so it provides benefits for bamboo farmers. Consequently, the plan has performed well so far. In addition to bamboo charcoal, bamboo farmers and manufacturers can obtain returns from selling the byproduct of "bamboo vinegar". Currently, there are around 40 kilns (including earthen and mechanized kilns) in Taiwan, and the production is estimated to approximately 250 ton yr^{-1}. In addition, to increase the competitive advantage in markets against cheaper imported bamboo charcoal and protect consumer safety, an official certification mechanism, called "CAS (Chinese Agricultural Standard)", to evaluate the bamboo charcoal quality of local manufacturers has been implemented for many years.

In the operation of the certification mechanism, several of checkup items are thoroughly inspected and appraised during document reviews and on-site evaluations for applicants of the CAS certification. These checkup items include the environment of the operation area, facilities hardware, raw materials, manufacturing processes, quality, environmental hygiene, and warehouse management. In addition, the labeling must carry important information such as the index of refinement (electrical resistivity), fixed carbon content, BET value (the specific surface area), hardness, species and

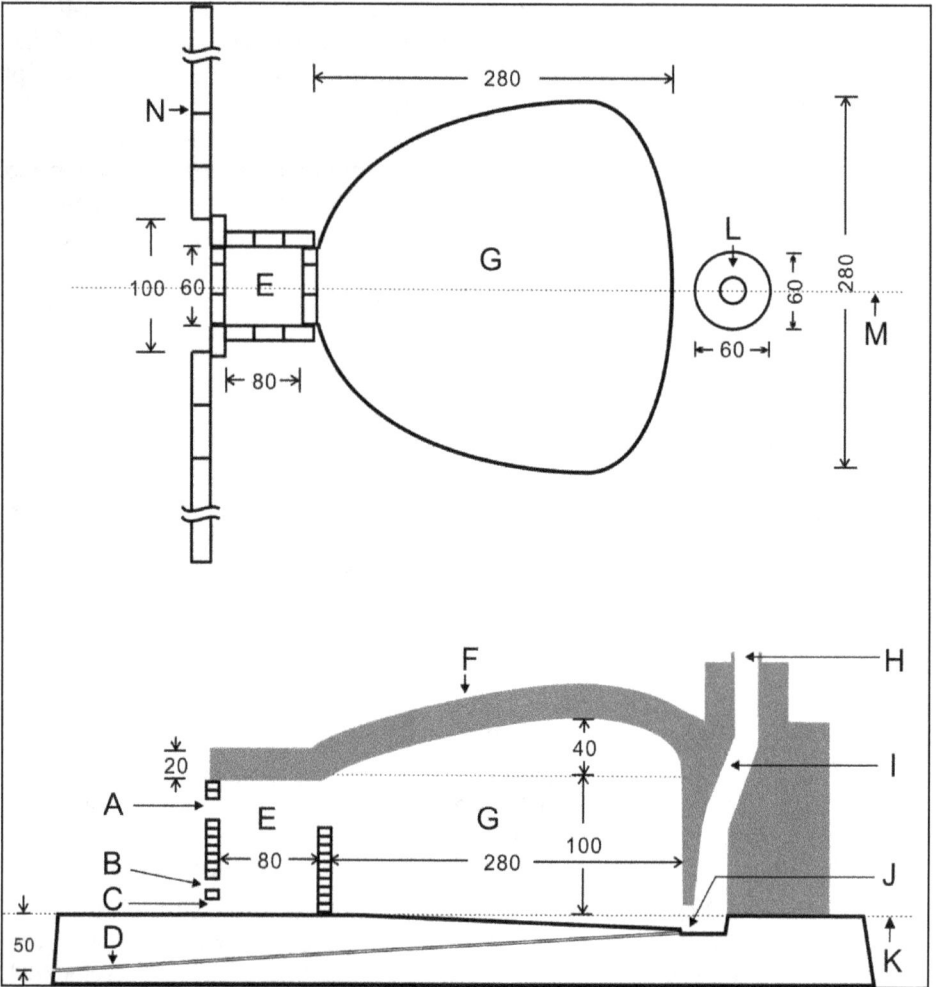

Figure 3.4: Top and Side Views and Dimensions of an Earthen Kiln.
A, Fuel Intake; B, air Intake; C, ash exit; D, scupper; E, fuel chamber; F, kiln top;
G, carbonization room; H, chimney; I, smoke channel; J, smoke exhaust hole;
K, horizontal baseline; L, chimney skirt; M, central line; N, revetment. All dimensions
are In centimeters (Lin *et al.,* 2004).

origin of the bamboo, manufacturing methods, shape, net weight, user instructions
and important notices, production date, manufacturer, and vendor (CAS 2006).

Prospects for the Bamboo Industry

To promote and revitalize the bamboo industry, Taiwanese agriculture authorities
and other relevant authorities gathered together and drew up the following strategies
and programs.

1. To Continue Developing Bamboo Charcoal Production

To extend the production scale of bamboo charcoal, mechanized kilns will be increasingly produced. Activated bamboo charcoal will be developed and new uses will be created, such as in the fields of soil improvements, food additives, heavy metal absorbents, materials for bio-medicine, etc.

2. To Create Higher Value-Added Fine Bamboo Craft Products

To further increase value-added fine bamboo craft products, it is necessary to select better-quality bamboo and cooperate with bamboo artists to develop exquisite and highly beneficial products during processing. At present, exquisite bamboo furniture and bamboo swords still possess development potential among bamboo products. Most remaining factories of bamboo craft production are located in Chu-shan Township, Nantou County, in central Taiwan. A specific processing park for the bamboo industry is still in operation there. These factories can be retained because they possess the advantages of process rationalization in production management through years of accumulated experience (Chang 2005).

3. To Develop the Bamboo Pyrolytic Oil for Bioenergy

Bamboo pyrolytic oil is produced by a rapid pyrolysis method, and this technology has moved beyond the laboratory scale. The potential contribution of pyrolytic oil production to substitute for fossil fuels as a part of boiler energy due to the abundance of raw materials has garnered more attention recently, and its development has been strengthened. This production is highly expected to enhance resource utilization of indigenous bamboo and make contributions in reducing emissions of carbon dioxide.

4. To Develop Bamboo Pellets for Bioenergy

Bamboo pellets are manufactured through a physical high-pressure treatment. The idea of bamboo pellets is derived from wood pellets. Because wood pellet manufacturing has advantages of less technical requirements, low capital investment, fewer environmental impacts, high market potential, and feasibility to develop local energy systems with pellet boilers, this product has high market potential in the world. Some studies showed that wood pellet consumption has grown from 2.7 million short tons in 2001 to 12.6 tons in 2008 on a global basis, which is an annual growth rate of about 25 per cent (Lin 2009). European countries consumed over 75 per cent of the pellets in 2008 what is driven by EU carbon and renewable energy policies. In Taiwan, there are few timber resources and waste woody materials with which to produce wood pellets, because of the policy prohibiting cutting in natural forests since 1990, yet bamboo resources are abundant and could be substituted for wood materials. The benefits of developing bamboo pellet production and its utilization in energy are expected not only for the similar benefits as bamboo pyrolytic oil production, but also creating job opportunities in farm communities. Therefore, to develop bamboo pellets instead of wood pellets will be an important aspect of bamboo utilization in Taiwan.

Conclusion

Bamboo is a significant forest product, and bamboo manufacturing is an essential traditional industry in Taiwan. However, the Taiwanese bamboo industry conspicuously declined as a result of changes in the industrial environment. To promote the amount of bamboo utilization, restore the bamboo-processing industry and bamboo production once again, revive the economic activities in villages, and recover the prosperity in farms and communities, Taiwanese agricultural authorities have successfully carried out a plan for bamboo charcoal production in abundant bamboo forest areas with an official certification mechanism of products. The related authorities have planned several beneficial projects as well. The prospects of reviving the bamboo industry in Taiwan can be expected if these projects are enhanced and to continuously implemented in the coming decades.

References

CAS. (2006). CAS categories and standards for woody products. Taipei, Taiwan: Chinese Agriculture Standard (CAS).

Chang, M. C. (2005). The development and adjustment strategies of bamboo craft industry of Chu-shan town, Nantou County [thesis]. Changhua, Taiwan: Department of Geography, National Chang-Hua University of Education. 113 p. [in Chinese]

INBAR (International Network for Bamboo and Rattan). (2009).The climate change challenge and bamboo- mitigation and adaptation 2009.

Lee, J. K. C., Lai, C. S., and Lien, C.C. (1993). An economic analysis of bamboo marketing in Taiwan. Quart. J. Exp. For. Nat. Taiwan Univ. 7(1):127-155. [in Chinese]

Liese, W. (1987). Research on bamboo. Wood Sci. Technol:21:189-209.

Lin, Y. C. (2004). The current status analysis and future development of Taiwan bamboo industry [thesis]. Taichung, Taiwan: Department of Forest, National Chung-Hsing University. 100 p. [in Chinese with English summary].

Lin, Y. J., Hwang, G. S., and Yu, H. Y. (2004). Cost analysis for the building of bamboo charcoal kiln. Quart. J. Chinese Forestry 37(2):195-204. [in Chinese with English summary]

Lin, Y. J. (2006). Study on the production and marketing structures and the strategies of bamboo charcoal industry in Taiwan [dissertation]. Taipei, Taiwan: School of Forestry and Resource Conservation, National Taiwan University. 169 p. [in Chinese with English summary].

Lin Y. J. (2009). A new favorite for bioenergy utilization – wood pellet. Forestry Research Newsletter 16(6):17-19. [in Chinese]

Lü, C. M. (2001). Cultivation and management of bamboo forests, TFRI Extension Series No.135. Taipei, Taiwan: Taiwan Forestry Research Institute (TFRI). 206 p. [in Chinese]

Scurlock, J. M. O., Dayton, D. C., and Hames, B. (2000). Bamboo: an overlooked biomass resource?, Biomass and bioenergy19:229-44.

Widenoja, R. (2007). Sub-Optimal equilibriums in the carbon forestry game: why bamboo should win, but will not. [thesis]. Medford, MA, USA. Fletcher School of Law and Diplomacy, Tufts University.104 p.

Chapter 4

Development of Bamboo Green Building Material and its Application

Min-Chyuan Yeh

National Pingtung University Science and Technology

Introduction

Bamboo is one of most important forest resources, covering 152,300 ha or 7.2 per cent of forest area in Taiwan (Forest Bureau, 1995). There are 75 bamboo species, including 17 indigenous species and 58 exotic species (Lu, 1994). The six major commercial bamboo species include *Phyllostachys makinoi* (makino bamboo), *P. pubescens* (moso bamboo), *Dendrocalamus latiflorus* (ma bamboo), *Bambusa oldhamii* (green bamboo), *B. stenostachya* (thorny bamboo), *B. dolichoclada* (long-shoot bamboo). Bamboo can be used in many applications in the daily life depending on its characteristics and quality. Thorny bamboo and long-shoot bamboo were mainly used for making containers for vegetable transportation forming a weaving industry, and making religious paper in pulping industry. Moso bamboo was mainly used for construction scaffoldings, chopsticks, toothpicks and window curtains. Makino bamboo is used as posts supporting plants in the farm and furniture manufacturing, while ma bamboo and green bamboo produce bamboo shoot as vegetable. Those bamboo species were also extensively served as building materials for beam, post and wall element for residential houses and farm. In recent years, those industries gradually lost the competition due to the rises of labor cost and also replaced by new materials such as steel and polymer. The construction scaffolding market is now dominated by steel framed products to replace bamboo products. Many bamboo business investments are switched to south Asian countries and mainland China and then imported for more profits, which in advance impacts the survivors seriously.

The industries face the challenge keeping in upgrading production technology and developing innovative products against those low end products.

As the global warming becomes an issue today, there are several issues that need to be considered in the development of innovative products for the forest product industries. We care about the sustainability of materials, conservation of species, friendly environment, ability of reuse, reduce and recycle, the requirement of green building, green construction materials, and safety (Wang, 2009). Bamboo materials are extensively used in the pavilion, bridge, roof truss, and post and beam structures for traditional residential houses. However, the flexibility on the structural space is always limited by the diameter of bamboo culm and the rigidity of structures. It is well-known that glued laminated timber features dispensed natural defects, dimensional stability, superior strength, and even quality through the manufacturing techniques and can be applied in the design projects of large space, long span, and curved structures. It solves the limitation of the structure size using solid wood materials and becomes an important building structural member for the modern wood-based architecture. In this study, the structural laminated bamboo member with high strength performance, dimensional stability, resistance to decay and moisture is developed to solve the utilization limitation of bamboo in structural application.

Assembly of Bamboo Glued Laminated Timber

Glued laminated timber comprises a number of bamboo strips or lamina stacked and pressed horizontally or vertically after applied with adhesive. A piece of flat bamboo strip is obtained by planning a split culm to remove curved portions on both surfaces forming a basic element of glued laminated timber. The thickness of bamboo wall and the diameter changes along the growth height of culms which will affect the width and thickness of lamina. For lamination, bamboo culm should have a thick wall and large diameter. Moso bamboo was harvested at the age of 4~5 years with on average diameter of 124 mm at the lower end of the culm (*Yeh et al.,* 2009). Bamboo strips were obtained by splitting a culm into 6 pieces and then planning them into the dimensions of 6 x 33 x 1850 mm after being treated with a 1 per cent boiling boron solution and being kiln-dried. A commercial resorcinol phenol formaldehyde adhesive (RPF, type AD500, Sports Leader, Taiwan) with a 52-58 per cent solids content and hardener (type H501) of paraformaldehyde in a 69-73 per cent solution were used for laminating the bamboo strips. Laminated bamboo members of 30 x 120 x 1850 mm were assembled with vertical or horizontal layers by stacking bamboo strips sequentially with the surface of the epidermal layer facing the same direction. Glue was applied at 250 g/m², and pressure application was 1.47 MPa.

Evaluation of Finger Joint

One of the advantage of laminated bamboo products is the member length can be extended in any size by jointing the structural elements end to end with the effective joints. Finger joint is the most extensive approach used for glulam manufacturing and can be applied to the laminating bamboo process. Some investigations are performed and the results are as follows:

(I) Effect of Bamboo Species and Growth Height Variation

The bending tests of bamboo glulam members laminated horizontally and jointed with 12-mm long finger formation were performed. The MOR of finger-jointed ma bamboo glulam members fabricated with the lamina from lower portions of bamboo culm showed about 20.5 per cent lower than those from middle and upper portions of bamboo calm (Table 4.1). Similar results were found in the cases of moso bamboo, which was 9.6 per cent lower in MOR for the glulam member fabricated with the lamina from the lower portion of bamboo culm. Overall, finger-jointed bamboo glulam members made from the middle and upper portions of culms had 14.1 and 15.1 per cent, respectively, higher in MOR than that of from lower portions of culms. It is noted that the strongest finger-jointed bamboo glulam member was found in the case of the lamina from middle portions of moso bamboo culm with the finger profile shown on the width face of the beam, 38.7 per cent higher than the lowest group which was fabricated with the lamina from lower portion of ma bamboo culm with the finger profile shown on the thickness face of the beam. In general, finger-jointed moso bamboo glulam showed 10.7 per cent higher in MOR than that of ma bamboo glulam. And, the strength of jointed moso bamboo glulam members with finger profile shown on the beam width face was 9.0 per cent higher than those shown on the thickness face, but no significant difference in the case of ma bamboo. The finger-jointed ma bamboo glulam members fabricated with the lamina from middle portions of bamboo culm showed a better bending MOE, 37.9 per cent higher than those from lower portions of bamboo culm. And, a 31.1 per cent higher in the bending MOE was obtained for moso bamboo groups. On the other hand, there was no significant difference in MOE between the finger-jointed members made from two bamboo species and two finger profile orientation.

Table 4.1: Strength of Finger-Jointed Glulam made from different Bamboo Species and Positions of Culm.

Treatment	MOR (MPa)	Specific Strength	MOE (GPa)
P-L-H-T-2[1]	67.4±8.7[CD]	92.4±11.0	9.04±0.39[F]
P-L-H-W-2	73.0±5.8[ABC]	100.1±8.0	9.61±0.37[EF]
P-M-H-T-2	71.3±10.2[BC]	99.1±14.2	13.17±1.38[A]
P-M-H-W-2	81.1±9.8[A]	112.7±13.5	11.28±1.59[CD]
P-U-H-T-2	75.8±9.5[ABC]	99.8±12.5	10.52±0.97[DE]
P-U-H-W-2	79.6±11.4[AB]	104.8±15.0	11.70±1.29[BCD]
D-L-H-T-2	58.5±9.1[E]	106.2±16.6	10.01±0.49[EF]
D-L-H-W-2	60.3±4.7[DE]	109.7±8.6	8.86±0.56[F]
D-M-H-T-2	69.0±4.6[C]	113.2±7.4	12.40±1.06[ABC]
D-M-H-W-2	74.2±5.4[ABC]	121.6±8.8	13.62±0.61[A]
D-U-H-T-2	70.6±6.7[BC]	103.8±9.8	12.77±1.57[AB]
D-U-H-W-2	72.2±5.9[ABC]	106.3±8.7	11.43±1.46[CD]

1: P: moso bamboo; D: ma bamboo; L: lower portion; M: middle portion; H: top portion; W: width surface; T: thickness surface; 2: 12-mm finger length.

(II) Effects of Finger Length and Lamination Orientation

The bending tests of bamboo glulam members laminated with the lamina from lower portions of moso bamboo were performed. The MOR of 15- and 18-mm long finger-jointed bamboo glulam members laminated vertically were 19.8 and 17.6 per cent, respectively, higher than those of 12-mm long finger-jointed group (Table 4.2). In the cases of members laminated horizontally, both 15- and 18-mm finger jointed bamboo members had MOR 13.5 per cent higher than the 12-mm finger-jointed group. It is noted that the strongest finger-jointed bamboo glulam member was found in the case of the 18-mm finger with the finger profile shown on the width face of the vertically laminated beam, 50.1 per cent higher than the lowest group which had 12-mm finger with the finger profile shown on the thickness face of the vertically laminated beam. Overall, finger-jointed bamboo glulam members laminated vertically showed a slight higher MOR, 7.7 per cent, than that of laminated horizontally. And, the bamboo glulam members jointed with the finger profile shown on the width face was found had better MOR, 17.6 per cent higher than the members with the finger profile shown on the thickness face of beam. The results indicated that the MOR of 12-, 15-, and 15-mm finger-jointed bamboo glulam members which were laminated vertically and finger profile shown on the width face were 21.0, 12.8, and 36.6 per cent, respectively, higher than those members with finger profile shown on the thickness face of the beam. Similar tendency was also found for the bamboo members laminated horizontally.

Table 4.2: Comparison of Static Bending Properties for Finger-Jointed Glulam Laminated in Vertical and Horizontal Direction and Jointed with various Finger Length.

Treatment	MOR (MPa)	MOE (GPa)
P-L-V-W-2[1]	80.3±7.4[CDE]	9.38±0.38[DE]
P-L-V-W-5	93.1±2.6[AB]	10.62±0.86[ABC]
P-L-V-W-8	99.6±12.0[A]	10.50±1.14[BCD]
P-L-V-T-2	66.3±4.3[G]	9.49±0.56[E]
P-L-V-T-5	82.5±7.7[CD]	10.69±1.43[ABC]
P-L-V-T-8	72.9±5.8[DEF]	11.74±0.93[A]
P-L-H-W-2	73.0±5.8[EFG]	9.61±0.37[CDE]
P-L-H-W-5	85.8±7.3[BC]	10.97±1.19[AB]
P-L-H-W-8	83.9±12.5[CD]	10.56±1.82[BC]
P-L-H-T-2	67.4±8.7[FG]	9.04±0.39[E]
P-L-H-T-5	73.3±14.0[EFG]	9.55±0.31[CDE]
P-L-H-T-8	75.8±7.7[EFG]	9.87±0.75[BCDE]

1: P: moso bamboo; L: lower portion; V: vertical lamination; H: horizontal lamination; 2, 5, 8: 12-, 15- 18-mm finger length.

For the MOE properties, the 15- and 18-mm finger-jointed bamboo members showed 11.5 and 13.7 per cent, respectively, higher than that of 12-mm finger-jointed groups. On the other hand, there was no significant difference in MOE between the finger-jointed members laminated in either vertical or horizontal directions and two finger profile orientation.

Mechanical Properties of Laminated Bamboo Beams

(I) Shearing, Compressive, and Bonding Strength

The bamboo strips were laminated for shearing, compressive and bonding tests. The laminated bamboo from middle and upper growth portions of culms exhibited higher compressive strength, 26.7 and 32.4 per cent, respectively, than that from lower growth portion of culm (Table 4.3). In general, the compressive strength of laminated moso bamboo showed 20.7 per cent higher than that of ma bamboo. Also, moso bamboo showed superior shearing strength to that of ma bamboo by 66.3 per cent. Although significant difference in shearing strength of laminated bamboo were found among different growth height of ma bamboo, moso bamboo from lower, middle and upper growth portions showed similar values. It is noticed that the bonding strength of laminated bamboo was only 57.3 and 43.5 per cent, respectively, of shearing strength of moso bamboo and ma bamboo showing a critical strength properties for making bamboo glulam products. The bonding strength of moso bamboo can meet the minimum requirement for making wood glulam member of southern pine and Douglas fir, while better gluing process may be needed for improving bonding performance for ma bamboo (CNS, 2002).

Table 4.3: Shearing and Compressive Strength of Laminated Bamboo.

Treatment	Density (g/cm³)	Shear Strength (MPa)	Compressive Strength (MPa)	Bonding Strength (MPa)
P-L[1]	0.73±0.01[B]	16.8±0.5[A]	54.2±2.9[B]	12.2±0.7[A]
P-M	0.72±0.01[B]	17.4±0.7[A]	66.1±1.9[A]	10.1±0.4[B]
P-U	0.76±0.01[A]	17.1±0.7[A]	69.6±2.7[A]	7.1±0.1[C]
D-L	0.55±0.03[E]	9.2±0.7[D]	42.5±2.9[C]	6.2±0.8[D]
D-M	0.61±0.03[D]	10.3±0.3[C]	56.4±5.4[B]	4.2±0.7[E]
D-U	0.68±0.02[C]	11.4±0.6[B]	58.4±4.7[B]	2.5±0.6[F]

1: Symbols is as listed in Table 4.1.

(II) Flexural Properties of Structural Laminated Bamboo Beams

The laminated beam is first assembled with a size of 30 by 120 by 2000 mm and then glue two pieces into 60 by 120 by 2000 mm member. Beams are also identified based on the materials produced from lower, middle and top portions of bamboo culm using moso bamboo and ma bamboo. The static bending MOR of beams made of moso bamboo is 17.7 per cent higher than that of ma bamboo while the MOE values of ma bamboo beams are 9.8 per cent higher than that of moso bamboo (Table 4.4). A tendency of increasing MOR is found as the beams laminated with bamboo laminae from top portion instead of lower and middle portions of culm for both ma bamboo (9.3~10.0 per cent) and moso bamboo (9.4~11.1 per cent). In the case of MOE, the structural laminated bamboo beams made of both moso and ma bamboo laminae from top portions are better than those from lower portions of culm, *i.e.*, 20.1 and 18.0 per cent, respectively.

Table 4.4: Flexural Properties of Structural Laminated Bamboo Beams.

Treatment	MOR (MPa)	MOE (GPa)	Equiv. Dist. Load (KN/m)
P-U[1]	111.7±2.7	11.18±0.39	4.59±0.63
P-M	101.2±4.4	11.96±0.49	5.17±0.61
P-L	99.3±6.8	9.51±0.49	4.04±0.39
D-U	91.5±4.5	13.43±0.93	5.97±0.46
D-M	82.4±4.1	11.67±0.39	5.27±0.51
D-L	83.0±2.1	10.98±0.59	4.66±0.22

1: Symbols is as listed in Table 4.1.

Composite Boxed Beam

Box beams are designed to efficiently use small sizes of wood-based materials combining the reinforcement of bamboo materials in structural applications. The outer surfaces of the top and bottom of a beam produce maximum compressive and tensile stresses when subjected to a flexural load. Consequently, the flexural performance of a box beam can be improved by using stronger flange members. The bamboo members were first finger-jointed longitudinally to a length of 3650 mm with RPF adhesives using a longitudinal finger jointer (model: KMFJ-400S, Chuan Chier Industrial, Taiwan). The finger length was 12 mm, the finger spacing was 4 mm, and tip width was 0.65 mm; it was processed using a finger shaper (model: KMFJ-400, Chuan Chier Industrial) The laminated bamboo members were used as flange material and Japanese cedar solid wood as web material to fabricate box beams of 120 mm wide and 150 mm deep. Four types of laminated bamboo/wood composite box hollow beams with bamboo strips oriented horizontally or vertically were fabricated with RPF adhesives, as shown in Figire 1. Two box hollow beams were further reinforced with either box nails or wood screws. Both fasteners were driven from upper and bottom flanges to the web at a 300-mm spacing during box hollow beam fabrication. In addition, box hollow beams fabricated with Japanese cedar wood flanges and solid wood beams with the same dimension in cross section as the box hollow beam were included as control groups.

(I) Flexural Properties of Laminated Bamboo Beams

The static bending tests were performed on the laminated bamboo with or without finger joint processing. The MOR of bamboo beams with strips laminated vertically was 19.1 per cent higher than that laminated horizontally as shown in Table 4.5. Both MOR values of laminated bamboo beams were much higher than those of moso bamboo in the original round shape, *i.e.*, 51.1 MPa, as reported by Yeh (1994). Similar results were found for the specific strength if the variation in the material density parameter was considered. In the case of laminated bamboo beams with finger joints,

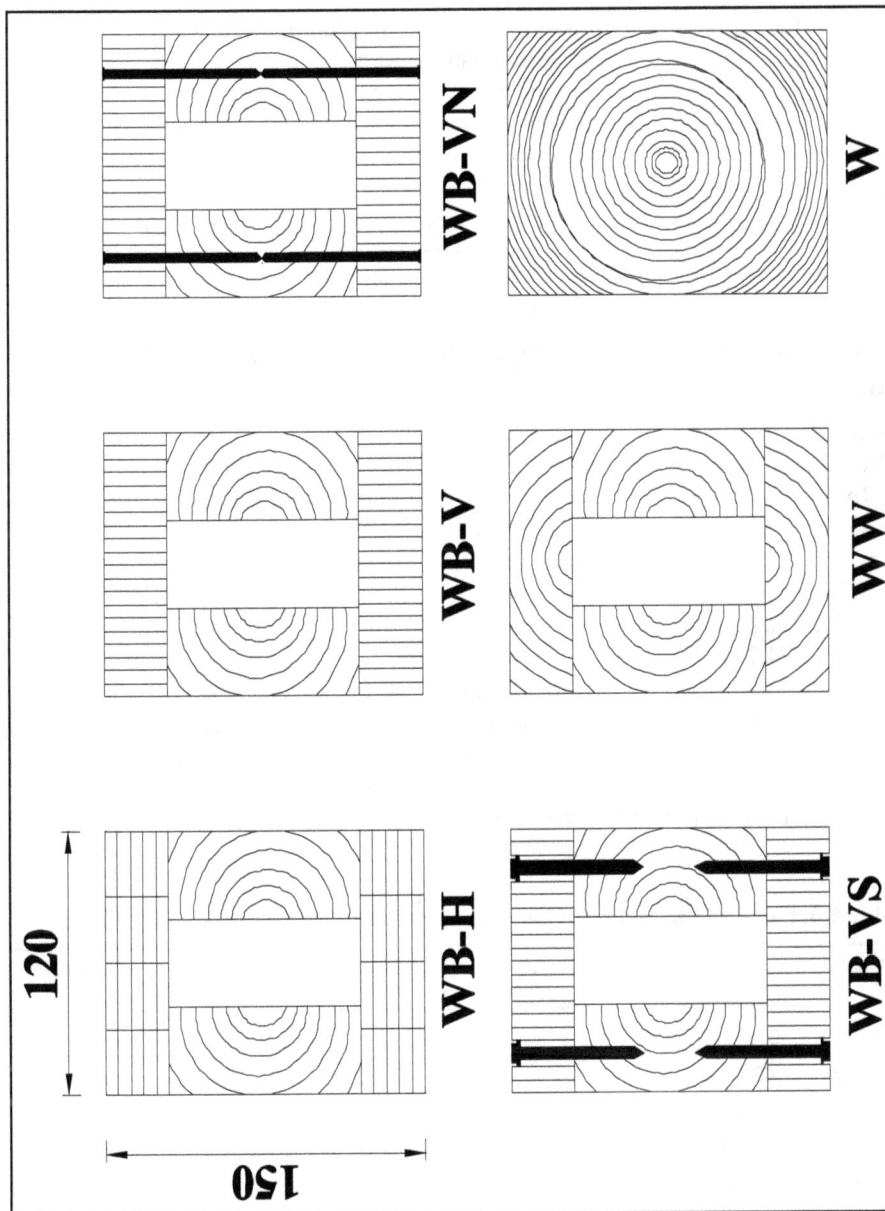

Figure 4.1: Types of Laminated Bamboo/Wood Box Hollow Beams.

Unit: mm (W, Japanese cedar; B, laminated bamboo; V, vertical; H, horizontal; N, box nail; S, wood screw)

the MOR values were significantly reduced due to failure at the critical joints. The efficiencies of the finger joints in the bending tests were 64.3 and 53.2 per cent for beams laminated horizontally and vertically, respectively, which showed similar performance compared to hardwood materials.

Table 4.5: Flexural Properties of Laminated Bamboo Beams with and without Finger Joints.

	B-H[1]	B-V	B-FH	B-FV
Modulus of rupture (MPa)	104.9 ± 14.7	124.9 ± 7.2	67.4 ± 8.7	66.4 ± 4.4
Specific strength (MPa)	143.7	171.0	92.4	90.9
Modulus of elasticity (GPa)	10.74 ± 1.15	11.25 ± 0.72	9.04 ± 0.39	9.21 ± 0.38

1) B, laminated bamboo; H, horizontal; V, vertical; F, finger-jointed.

(II) Flexural Properties of Laminated Bamboo Box/Wood Hollow Beams

The maximum bending capacities of box hollow beams using laminated bamboo members either horizontally or vertically as flanges were 31.3 and 49.7 per cent, respectively, higher than those of beams with Japanese cedar flanges, as shown in Table 4.6. The bending MOR of small Japanese cedar clear wood is about 55.9~71.2 MPa, which results in 47.3~75.4 per cent higher magnitudes for laminated bamboo beams. This explains the advantage of using laminated bamboo lumbers instead of low-quality plantation lumber for flanges. Improved maximum bending capacities, *i.e.*, 69.1 and 74.2 per cent, were found for laminated bamboo/wood box hollow beams further respectively reinforced with 10 d box nail and wood screws on the flanges. Furthermore, the maximum bending capacities woo box hollow beams were only slightly less than those of Japanese cedar solid wood beams, *i.e.*, 7.7 per cent, and laminated bamboo/wood composite box hollow beams also showed better flexural performance than solid wood beams.

Table 4.6: Flexural Properties of Laminated Bamboo/Wood Box Hollow Beams.

Composite Beam Type	Max. Load (kN)	Bending Moment (kN-m)	Modulus of Rupture (MPa)	Apparent Modulus of Elasticity (GPa)
WB-H[1]	37.06 ± 1.74[bc2]	21.68 ± 1.02[bc]	49.21 ± 2.24[b]	9.72 ± 0.31[a]
WB-V	42.28 ± 1.66[ab]	24.73 ± 0.97[ab]	53.81 ± 3.12[ab]	9.91 ± 0.43[a]
WB-VN	47.74 ± 3.96[a]	27.93 ± 2.32[a]	61.69 ± 5.19[a]	10.40 ± 0.10[a]
WB-VS	49.18 ± 4.86[a]	28.77 ± 2.94[a]	62.70 ± 6.01[a]	10.24 ± 0.24[a]
WW	28.23 ± 7.12[d]	16.52 ± 4.17[d]	33.59 ± 8.40[c]	7.59 ± 0.01[b]
W	30.59 ± 0.51[cd]	17.89 ± 0.30[cd]	36.89 ± 0.53[c]	6.65 ± 0.63[c]

1) W, Japanese cedar; B, laminated bamboo; H, horizontal; V, vertical; N, box nail; S, wood screw.

2) Means (superscripts a, b, c, and d) within a given column with the same letter do not significantly ($\alpha \geq 0.05$) differ as determined by Duncan's multiple-range test.

The moduli of rupture of box hollow beams using vertically laminated bamboo members as flanges were 60.2~86.6 per cent higher than those wood box hollow beams (Table 4.6). The MOR of beams with horizontally laminated flanges was also 46.5 per cent higher than that of wood box beams. All laminated bamboo/wood composite box hollow beams showed apparent MOE values 28.1~37.0 per cent higher than those of solid wood box hollow beams. Furthermore, the apparent MOE of the wood beam was 14.1 per cent higher than that of solid beams.

(III) Design Considerations

According to the design criteria for wood-framed residential construction, the allowable bending deflection limitation of a beam member should be less than 1/300 of the span (Ministry of the Interior, 2003). The loading capacities of the bamboo/wood composite box beams under the limitation of the design deflection are shown in Table 4.7, which indicates about 111.0~15.4 per cent of the maximum bending loads. The equivalent uniformly distributed loads of bamboo/wood box hollow beams were 30.9~34.9 per cent higher than that of the Japanese cedar solid beam. Improvements in the bending capacities of 40.9 and 47.5 per cent were found for beams further reinforced with nails and wood screws on the flanges compared to solid wood beams. This verifies the improvement structural performance of laminated bamboo/wood composite box hollow beams.

Application: Design and Construction of a Bamboo Pavilion

The case was selected to build a bamboo pavilion at the campus of an elementary school in Chu-Shan, which is the major bamboo priority region and has many related manufacturing plants and business. The landscape was designed to match the idea for the purpose of biological education and recreation of children and local citizens. A shed and a bridge were designed in this case.

(I) Foundation

The dimension of pavilion is 3500 (width) by 2300 (depth) mm on the poured concrete platform with 390 mm in height. The structure was surrounded by pathway covered with pedestrian bricks 50 mm above ground. The H type stainless steel fasteners for posts were installed on the platform during the operation of pouring concrete. The stepped concrete foundation for bridge butts is 1500 mm wide 390 mm high and reinforced by steel bars.

(II) Main Structure

The height of pavilion is 3600mm with a projected roof size of 3060 by 3600 mm. Four main posts use laminated bamboo members with 120 by 120 mm cross section (Figure 4.2). The size of main beams is 90 by 120 mm and is decorated with the small diameter bamboo like swollen bamboo and Formosa bamboo above the beams, which are treated with preservatives. The L type stainless steel fasteners are used for the connection between post and beam members. The roof structure is designed with a truss system providing light weight and rigidity. The roof pitch is 30 degrees and truss members of 36 by 90 mm laminated bamboo are assembled with nails and metal

Table 4.7: Loading Capacities of Laminated Bamboo/Wood Composites Box Hollow Beams under the Limitation of the Design Flexural Deflection.

Beam Type[2]	WB-H	WB-V	WB-VN	WB-VS	WW	W
Load (kN)	4.87 ± 0.09 (12.7 per cent)[1]	5.02 ± 0.14 (11.9 per cent)	5.25 ± 0.07 (11.0 per cent)	5.49 ± 0.07 (11.2 per cent)	4.34 ± 0.08 (15.4 per cent)	3.72 ± 0.10 (12.2 per cent)
Equivalent uniformly distributed load (kN m^{-1})	1.85 ± 0.03	1.91 ± 0.05	1.99 ± 0.03	2.09 ± 0.03	1.65 ± 0.03	1.41 ± 0.04

1) Calculated based on the ultimate bending load.

2) Beam types are described in the footnotes to Table 4.6.

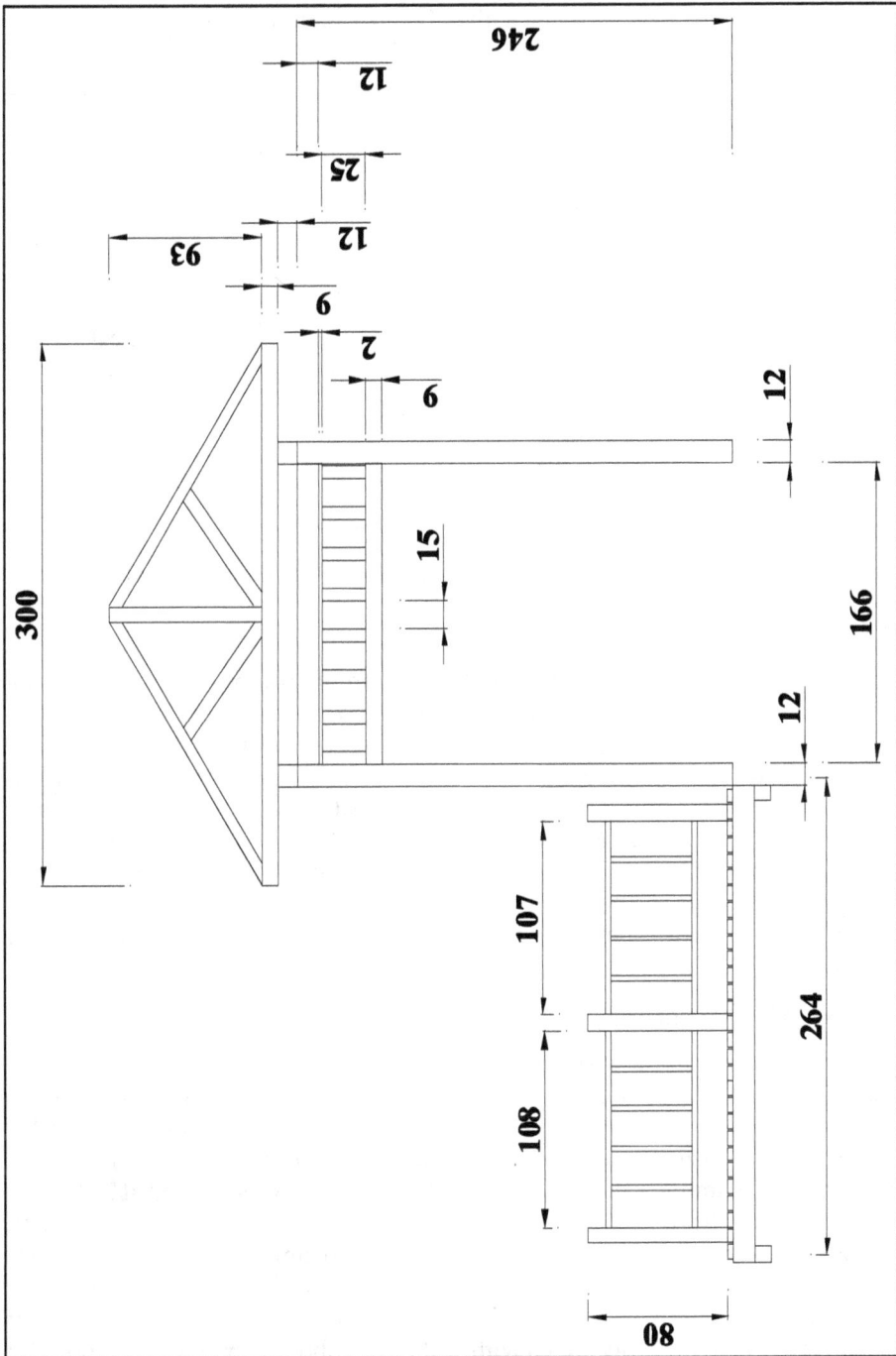

Figure 4.2: Side View of Pavilion Design Using Laminated Bamboo Members. (unit: cm)

plates (Figure 4.3). The pilot holes are drilled before preservation process. The span of main beams is 3000 mm and the spacing between each truss is 600 mm. Each truss is installed on the main beam member with galvanized steel angles by galvanized 4d box nails.

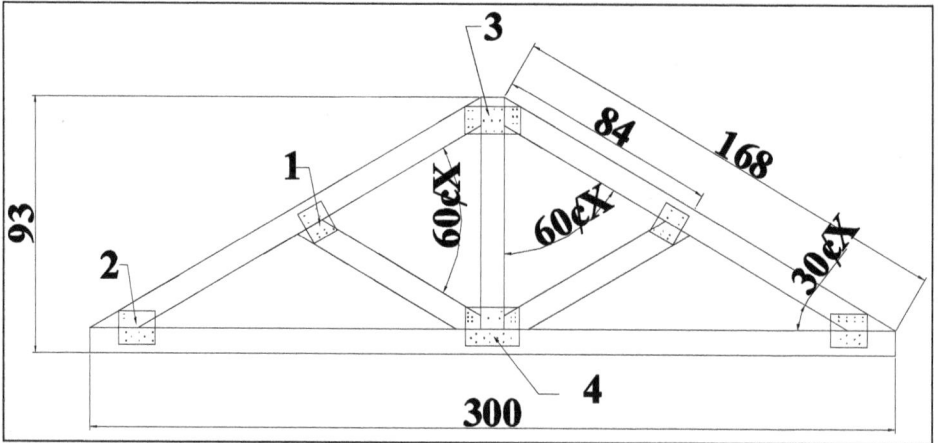

Figure 4.3: Roof Truss System Assembled with Nail Plate. (unit: cm).

(III) Pedestrian Bridge

The bridge is over a small stream with a span of 2640 mm and seats on the stepped reinforced concrete foundation on the both sides of bank. The sills use treated laminated bamboo members with the size of 90 by 90 mm and are installed with bolts on the foundation. Four laminated bamboo members with the size of 60 by 120 mm are used as the girder of bridge. The width of bridge pathway is 1170 mm and covered with laminated bamboo flooring with the size of 30 by 80 by 1170 mm for each member, which are fastened with wood crews. The size of laminated bamboo posts are 90 by 90 by 800 mm on the both sides of pathway and are fixed with mechanical fasteners on the flooring. The size of railings is 30 by 90 mm and spacing of baluster is 190 mm.

(IV) Roof Construction

The roof truss system is covered with water resistant plywood with 15 mm thickness as sheathing panels. An asphalt rolls is covered on the plywood as underlayment. Shingles are 320 by 320 mm in size, which uses water resistant 15-mm plywood as a base material after preservative treatment and then overlaid with treated an 1-mm bamboo veneer using RF adhesives. Glue was applied at 200 g/m², and pressure application was 0.98 MPa. Bamboo shingles are installed with galvanized 6d box nails and overlap each other for a 120 mm exposure length (Figure 4.4). Roof ridge is covered with 4-m long moso bamboo culm having 150-mm diameter after preservative treatment. The size of bamboo eave member or fascia is 36 by 80 by 3720 mm. Interior of roof is decorated with makino bamboo, i.e. 50-mm bamboo culm cut into half and treated with preservatives.

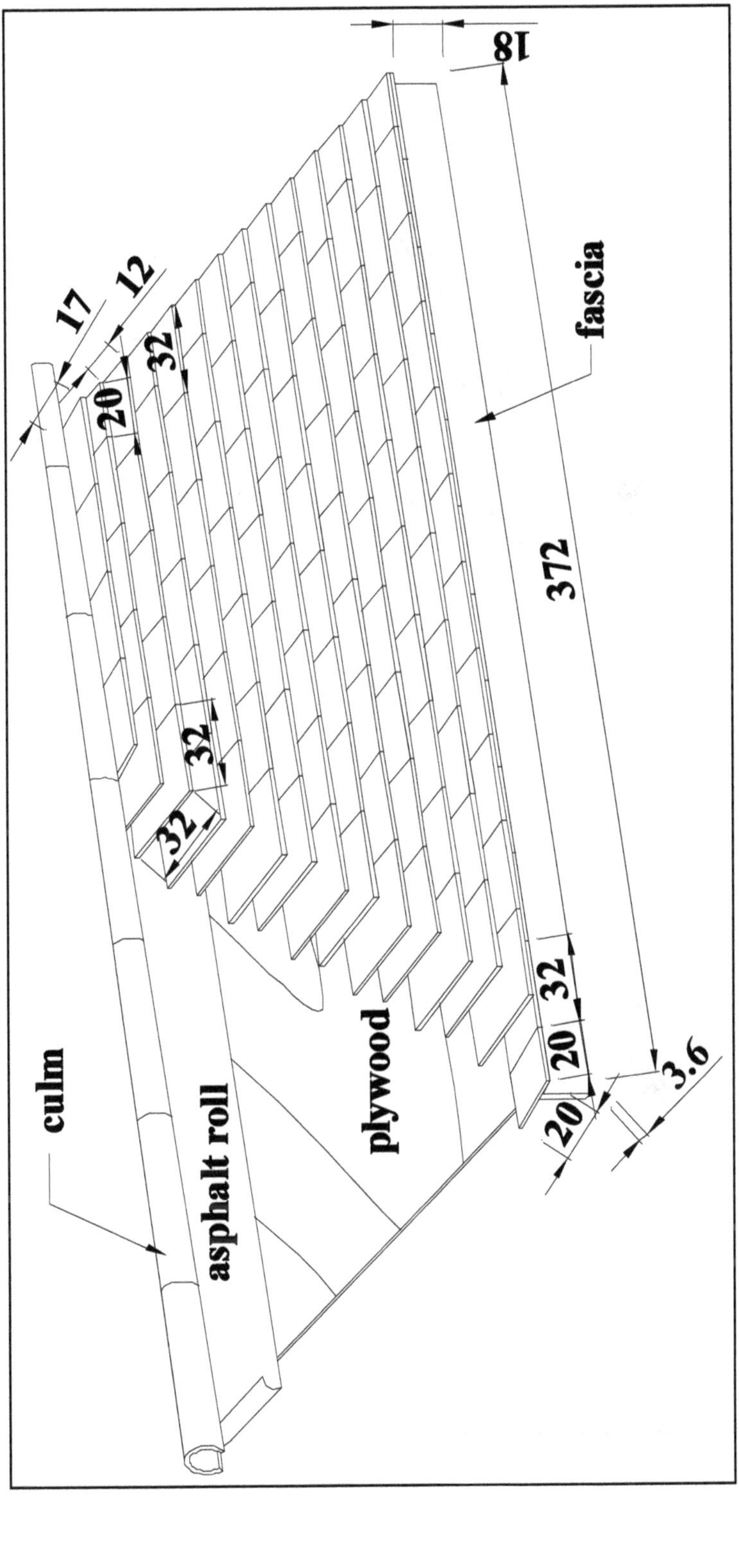

Figure 4.4: Structures of Roof Design for Bamboo Pavilion. (unit: cm)

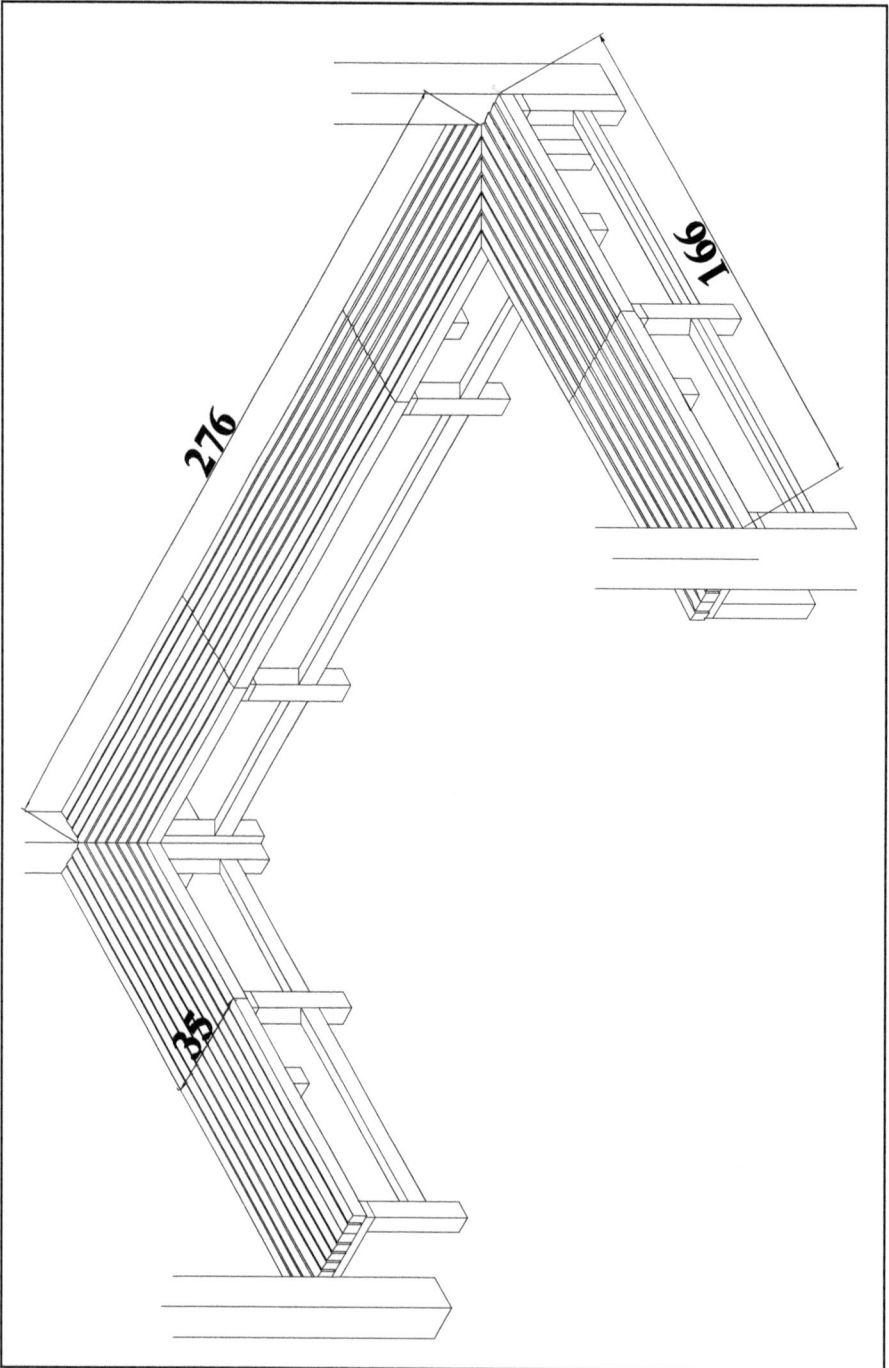

Figure 4.5: Chair Designed with Laminated Bamboo Products. (unit: cm)

(V) Chairs

The chairs are designed for the children in elementary school. The height of chairs is 398 mm and 330 mm wide, 1660 mm long along three sides of pavilion except the entrance (Figure 4.5). The connection between column and rail is mortise and tenon joint reinforced with steel angles. Laminated bamboo members are assembled on the top of chair by wood screws.

Overall sketch of pavilion structures is shown in Figure 4.6. It is suggested that all the fasteners shall be durable and anti-rustic. Major connections of the structures can use stainless steel fasteners such as H type fasteners at the post base and L type fasteners at the connections between post and beam members. A tendency of split problem at the end of bamboo members shall be considered, and enlarge the end distance and spacing for multi-bolted member is suggested. It is appropriate for a longer nailing spacing and pilot hole as assembling truss system with metal plate.

Conclusion and Suggestions

Structural laminated bamboo products can be a value-added building material because it features high strength, fine texture, and better dimensional stability. It is suitable for high quality structural applications. Certain factors shall be considered during the manufacturing in case of commercial production.

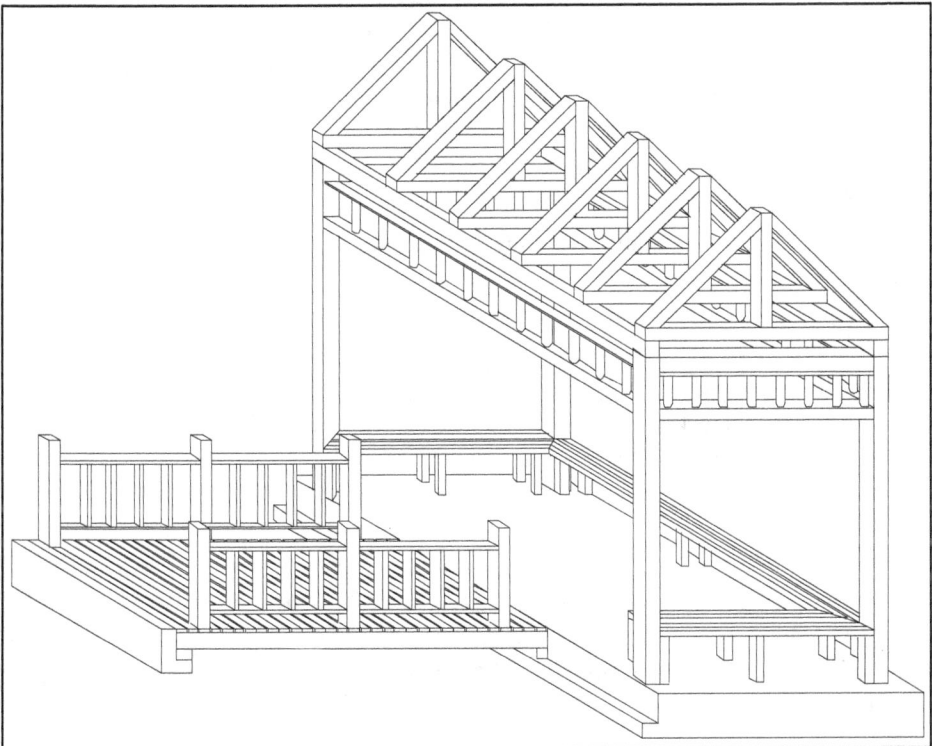

Figure 4.6: Overall View for Pavilion Design. (unit: cm)

1. For the purpose of increasing the processing yields, the relation between culm size and the thickness, width, and length of bamboo strips shall be well understood in the selection of raw materials. The adjustment of processing techniques and the acceptance on the market for changing various bamboo species shall be investigated.

2. To prevent molds and deformation of bamboo strips, moisture control is a critical procedure during the processing of bamboo splitting, planning, and laminating steps.

3. To raise the production capacity and lower the cost, adequate production equipment and production management skills are required to improve the efficiency of glue operation, reduction of labor, and saving in processing time.

4. To increase the utilization yield, the losses of bamboo materials in the planning, finger-jointing, and laminating procedures on the processing line shall be considered.

Practicality and convenience for design works are priorities in the engineering application. Some suggestions are made for the future investigation:

1. Effect of finger jointing techniques on the mechanical properties of laminated bamboo products.

2. Establishing the allowable stress data for structural members of laminated bamboo as the reference for structural analysis and design projects.

3. Evaluation on the long-term weathering, durability, and resistance on insects for structural laminated bamboo products and the efficiency of preservative treatment.

References

CNS. (2006). Structural glulam. Standard Investigation Bureau, ROC. CNS No.11031, 34.

Forestry Bureau. (1995). The third forest resources and land use inventory in Taiwan. Council of Agriculture, Forestry Bureau, Taiwan ROC, pp. 258.

Lu, C. M. (1994). Resources and utilization of bamboos in Taiwan. Proceedings of an international workshop on problem analysis on bamboo research in South East Asia. Edited by Kung-Fu Liao and Ying-Shen Wang. COA, The Forest Products Association of ROC, Asia Agricultural Technical Service Center. Taipei, Taiwan, pp. 97-114.

Ministry of the Interior, (2003). Specification of wood-framed structure design and construction techniques. Taipei, Taiwan. Construction Magazine, pp. 5-1~24.

Wang, S. Y. (2009). New trend of wood-based materials corresponding to environment, conservation, sustainability, regeneration and safety requirement. Wood Architecture, pp. 1-20.

Yeh, M. C., Hong, W. C., and Lin, Y. L. (2009). Flexural properties of structural laminated bamboo/solid wood composite box hollow beams. *Taiwan J For Sci* 24, 41-49.

Yeh, M. C. (1994). Evaluation on the mechanical properties of moso bamboo. *Q J Chinese Forestry* 27, 107-118.

Chapter 5

Bamboos of Western Himalaya and Flowering Event

H.B. Naithani

Consultant, Botany Division, Forest Research Institute, Dehradun, Uttarakhand

The Western Himalaya is one of the well defined and better known phytogeographical region of Indian subcontinent. In his Sketch of the Flora of British India, Hooker recognized the western Himalaya botanical province as extending from Kumaun (presently in Uttarakhand) to Chitral (present in occupied Kashmir). On the other hand at present the states which falls under western Himalaya are Jammu and Kashmir, Himachal Pradesh and Uttarakhand.

The bamboo species found wild in the Western Himalaya are *Ampelocalamus patellaris* (Gamble) Stapleton, *Dendrocalamus strictus* (Roxb.) Nees, *Sinarundinaria anceps* (Mitf.) Chao and Renv., *Sinarundinaria falcata* (Nees) Chao and Renv., *Thamnocalamus falconeri* (Hook. f. ex Munro) Benth. and *T. spathiflorus* (Trin.) Munro. Apart from these, a few bamboo species are planted *viz. Bambusa bambos* (Linn.) Voss, *Bambusa nutans* Wall. ex Munro, *Dendrocalamus somdevai* Naithani, and *Phyllostachys aurea* Carr.

The study of bamboo flowering is a bedelevilled problem for the foresters and Scientists. There are three types of flowering reported in the bamboos. (i) Annual (ii) Sporadic (iii) gregarious. During the process of gregarious flowering each species/clump of bamboo governed by certain genetic mechanism related with dates or age of genetic stock, entire species at appropriate time switches over from vegetative growth to the production of flowers. It is also comparable to the working of biological – clock.

If the flowering signal generated at particular time, all population of a given species raised from the same seed source, no matter where they are established, would start flowering at the same period. It is fortunate that no incident of rodents during bamboo flowering has been reported from western Himalaya.

Ampleocalamus patellaris (Gamble) Stapleton: Although it is commonly found in eastern Himalaya, in western Himalaya it is confined only to one place *i.e.* Kolona near Naintal, Uttarakhand. It has no record of flowering from western Himalaya.

Dendrocalamus strictus (Roxb.) Nees: It is found in the foothills of Jammu, Himachal Pradesh and Uttarakhand. It has many gregarious and sporadic records from Uttarakhand. Gregarious flowering records are in Uttarakhand 1880, 1908 – 1913, 1909 – 1910, 1948, 1968, 1989, 1998 and 2009.

Sinarundinaria anceps (Mitf.) Chao and Renv.: This bamboo is confined to Uttarakhand and western Nepal, also cultivated in Britain and U.S.A. In 1910 and 1966 it gregariously flowered in England and in 1978 at Chamoli Garhwal, Uttarakhand and recently in 2008 – 2009 in Bageshwar district, Uttarakhand. It has a flowering cycle of 45-55 years.

Sinarundinaria falcata (Nees) Chao and Renv.: This species has a wide range of distribution. It is found from Jammu and Kashmir (Peer Panjal Ranges) to Arunachal Pradesh. It has many gregarious flowering records from Uttarakhand in recent past *i.e.* 1998, 2000, 2004 – 2006. It also gregariously flowered at Himachal Pradesh in 2002. It has a flowering periodity of 28-32 years.

Thamnocalamus falconeri (Hook. f. ex Munro) Benth.: The species has distribution from Uttarakhand eastward to Sikkim. It has may gregarious flowering records from Uttarakhand *i.e.* 1846, 1847, 1885, 1906, 1913 – 1914, 1982. In recent past it has gregariously flowered in 2002. It has flowering periodity of 28 – 33 years. Outside India it has flowering records in all over Europe and in Algeria during 1875 – 1876 and 1964 – 1969.

Thamnocalamus spathiflorus (Trin.) Munro.: Distribution from Uttarakhand eastward to Arunachal Pradesh. It has flowering records from Uttarakhand, *i.e.* in the years 1818 – 1821, 1863 – 1886, 1882, 1892-1893, 1898-1899, 1902, 1942 and in recent past in 2001 – 2002. Its flowering periodity is 60 years.

Bambusa bambos (Linn.) Voss.: In western Himalaya it is found under cultivation. It has gregarious flowering records from Uttarakhand in the years 1836, 1881, 1925 and in recent past in 1991-1992.

Bambusa nutans Wall. ex Munro.: In Western Himalaya, it is distributed in Himachal Pradesh and Uttarakhand. This bamboo flowers of and on sporadically as well as gregariously at long intervals. From Uttarakhand it has flowering records from Dehra Dun *i.e.* during 1840, 1893 – 1894, and 1980.

Dendrocalmus somdevai Naithani.: It is endemic to Western Himalaya and so far known only form Himachal Pradesh and Uttarakhand. It has no record of gregarious flowering. It sporadically flowered in 1991 in Uttarakhand and in 2002 in Uttarakhand and Himachal Pradesh.

Phyllostachys aurea Carr.: It is a native of China commonly planted in Himachal Pradesh and Uttarakhand hills at an altitude of 1500 m. It has no flowering record from Western Himalaya.

Chapter 6

Flowering Characteristics and Seedling Growth of Four Bamboo Species in Taiwan

Tsai-Huei Chen

Taiwan Forestry Research Institute

Introduction

Among the plants reproduced by clonal system, it is quite rare for the case of bamboo which have a long-life cycle, a single reproduction, gregarious flowering and death. The gregarious flowering of *Melocanna baccifera* is still a mystery, and the mechanism of bamboo flowering is being researched worldwide.

Therefore, researches on bamboo flowering, fruits, seedling growth, and the restoration of population structure are quite few. It took 20 years for the recovery of the previous bamboo structure in Sasa, Japan. However, for *Pseudosasa usawai* in Taiwan, even 10 years after flowering death, the stand was unable to restore the pre-flowering population (Lin, 1974).

In Hokkaido, Japan and China, following bamboo flowering, a massive growth in rodent population is observed owing to the availability of seeds from flowering (Janzen, 1976). Consequently, the period from seed fall to the early stage of seedling growth is most crucial in the bamboo's life cycle. Because bamboos take long time to flower, uncertainies involved with flowering events and subsequent death of the plants, it is difficult to get materials for research.

While theoretically, it is true that the genetic diversity of bamboo species can be expanded through the sexual propagation by flowering, it is still investigated for confirmation.

The death of bamboo after gregarious flowering affects the surrounding ecosystem. The panda in China, for instance, have been threatened by the gregarious flowering in *Bashania* bamboo. The serious impacts on ecosystem are also found for the gregarious flowering in *Melocanna baccifera* in India and Bangladesh (Jaksic and Lima, 2003).

The purpose of this paper is to investigate the flowering and regeneration situations of four bamboo species in Taiwan.

Flowering Characteristics of Bamboo

1. Bamboo in Tropical and Subtropical Zones

Clark (1997) claimed that plants of Poaceae were originated in tropical zone and expanded gradually to the temperate zone, therefore, bamboos in tropical and subtropical area are different in terms of morphology, life pattern, sustenance growth, and flowering (Makita, 1998). McClure (1966) pointed out that life pattern of bamboo can be recognized by the rhizomes. Basically, there are two types of bamboo rhizomes: Leptomorph rhizomes and Pachymorph rhizomes (Figure 6.1). Leptomorph bamboos are characterized by running rhizomes with culms coming from buds on the rhizomes. Pachymorph bamboos are portrayed by clustered culms coming from the head of rhizomes. In addition to these two basic types, an intermediate type is also available.

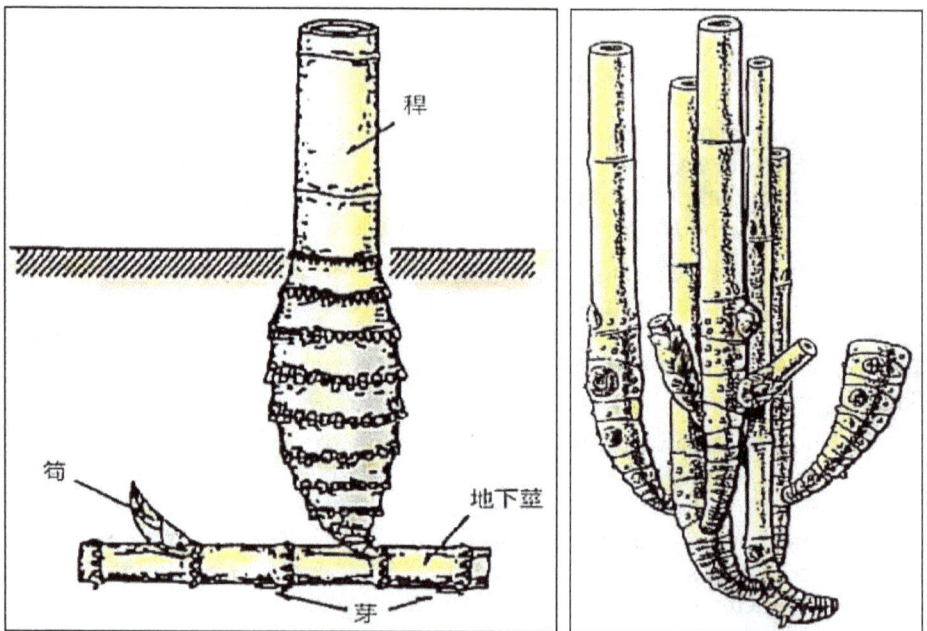

Figure 6.1: Leptomorph Bamboos (left) and Pachymorph Rhizomes (right).

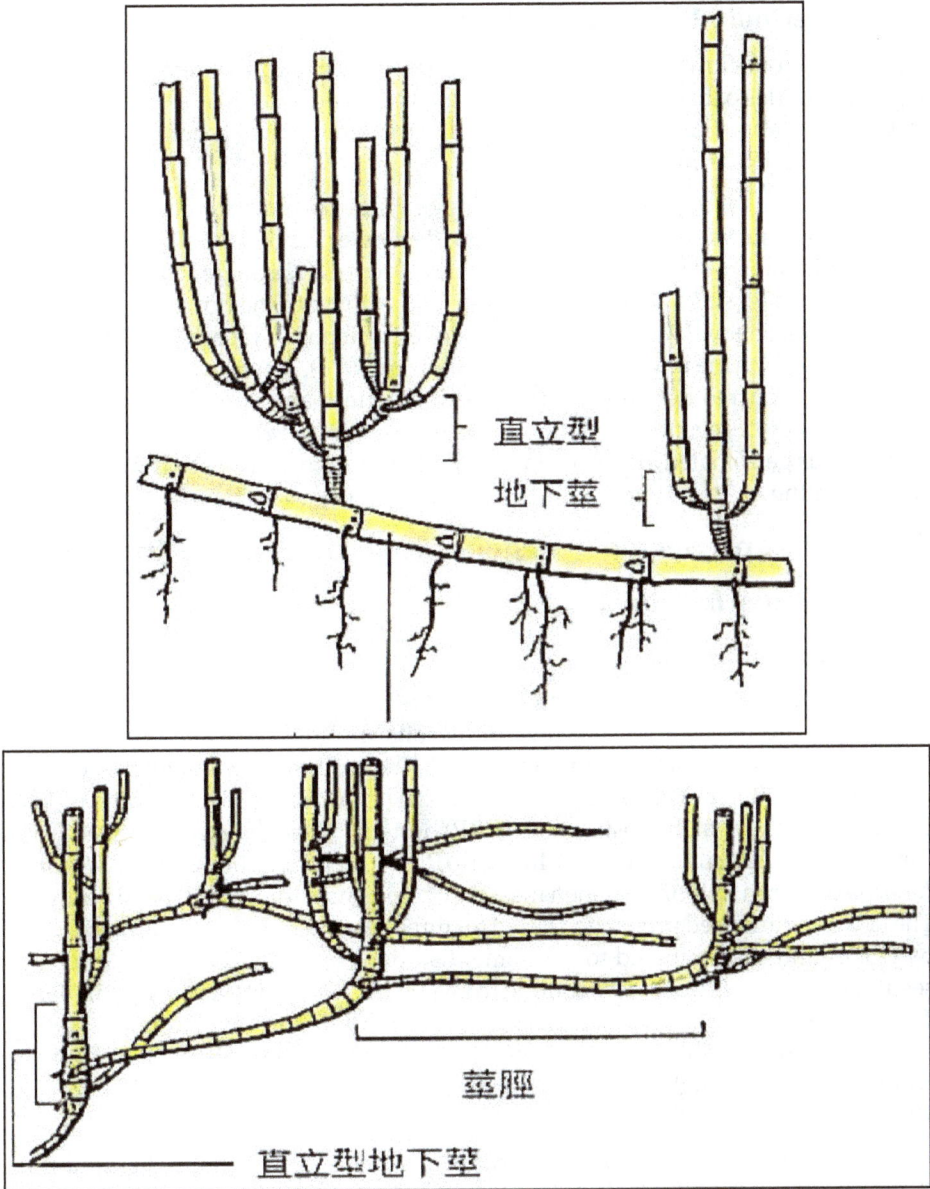

Figure 6.2: Matamoph ÉI Bamboo (left) and Matamoph I of Melocanna (right).

In Taiwan, based on Lin's study (1976), the intermediate type can be further classified into two types: Matamoph ÉI and Matamoph I of Melocanna (Figure 6.2).

2. The Periodic Flowering in Bamboo

The bamboo flowering events were reported since long. Owing to the existence of great varieties of bamboos, the flowering patterns are also diverse. However, the pattern of bamboo flowering can be basically divided into two types *i.e.*, gregarious and sporadic flowering.

In addition to the gregarious flowering of *Melocanna baccifera* occurred in India and Bangladesh, the large scale nation-wide flowering of madake bamboo was observed in Japan in 1970 and it lasted for 10 years (Kasahara, 1971). In Japan, *Phyllostachys pubescens* were blossomed sporadically with the flowering interval of 67 years (Watanabe, 1982). However, madake bamboos in Japan were blossomed gregariously with the interval of 120 years (Janzen, 1976). Between these two species, differences occurred not only in the flowering pattern and periods of interval, but also in the way of regeneration after flowering. In *Phyllostachys pubescens*, 40-80 per cent of seeds germinated, however, in madake bamboo only a few seeds were capable of germinating.

Flowering and Seedling Regeneration of Bamboo in Taiwan

1. *Melocanna baccifera*

(1) History of Planting

Melocanna baccifera was originated in India, Bangladesh and Myanmar. Owing to the long node, high quality and numerous usages, *Melocanna baccifera* was introduced to many places around the world. The earliest record of introduction of *Melocanna baccifera* in Taiwan can be traced back to August, 1960. Mr. Yang of Joint Commission on Rural Reconstruction (JCRR) brought 53 seedlings from United States Department of Agriculture (USDA). In Puerto Rico, a gregarious flowering of *Melocanna baccifera* occurred in 1957, and then massive of seeds were produced in the next year. The flowering and seed production lasted for three years. Seeds produced was collected by USDA and disseminated to experiment stations of many countries. Among 53 seedlings of *Melocanna baccifera* coming from USDA, 18 seedlings were distributed to Taiwan Forestry Research Institute and planted at Liu-Kuei experimental station. However, only 2 seedlings survived after planting. Descendants distributed in Taiwan were produced through agamogenesis propagation from these two seedlings (Lin, 1974; Lu and Chen, 2009).

The planting of *Melocanna baccifera* in Taiwan begun in 1970. In Chung-Pu Experimental Station, TFRI, *Melocanna baccifera* was planted in May, 1970 on an area of 1.7 ha, and followed by Linwhachin Experimental Station, TFRI, in 1971 with 600m^2 in area (Lu and Chen, 2009).

(2) The Survey of Flowering and Seeds

Recently, the flowering in *Melocanna baccifera* was observed on November, 2007 in Chung-Pu Experimental Station, TFRI, and it was blossomed in February, 2008. The seeds were produced in April, 2008 with fruits 4-7 cm in length and 2cm in diameter. However, the flowering of *Melocanna baccifera* in Linwhachin Experimental Station, TFRI begun in April, 2008, the latest flowering in Taiwan. The shape of fruits indicated that the fruits collected from Taipei Botanical Garden were thin and small in size. However, the fruits were oval and big in Nanto Bamboo Specimen Garden. The fruits collected from Chung-Pu, Linwhachin and Shan-ping were of both types (Lu and Chen, 2009).

(3) Cycle of Flowering

The cycle of flowering in *Melocanna baccifera* varied from 30 years (Gamble, 1896) to 60 years (McCulre, 1966). Thanchunga (2004) reported that in Mizoram, a gregarious flowering of *Melocanna baccifera* occurred in 1911-1912, and again in 1959-1960. It was forecasted that the next flowering which occur in 2006-2007, which was confirmed by Shibata (2010).

The seedlings of *Melocanna baccifera* were introduced to Taiwan in 1960. The seedlings originated from individuals those blossomed in the year immediately before 1960. After 48 years of agamogenesis propagation, gregarious flowering of *Melocanna baccifera* was observed in Taiwan with an interval of three months in flowering between Chung-Pu Experimental Station and Linwhachin Experimental Station. This was attributed to the temperature difference between the two places.

(4) Natural Regeneration

Melocanna baccifera in Taiwan blossomed in 2008 starting from February to August. Unlike most Poaceae, the process of flowering, fruiting and mortality in *Melocanna baccifera* spanned for 6 months. The comparisons of seedling growth in Chung-Pu and Linwhachin Experimental Station are given below (Lu and Chen, 2009):

a. In Chung-Pu area, the flowering began in the early February, 2008. However, in Linwhachin area, the flowering began on 28th April, 2008, three months later than Chung-Pu area.

b. The *Melocanna baccifera* bamboo in Chung-Pu area had two types of seeds, but in Linwhachin area only one type of seed was observed.

c. The height of naturally regenerated seedlings was more than 4m in Chung-Pu area. However, in Linwhachin area, the seedlings were of 1-2 m in height.

d. The vines threatened bamboo seedling growth in both areas.

2. *Pseudosasa usawai*

(1) Distribution

Pseudosasa usawai is an endemic species of Taiwan. It was mainly distributed in low elevation mountain area in the northern, central and eastern Taiwan with the elevations from 80 to 1200m. Most *Pseudosasa usawai* are located in the Yangminsan national park with the area of 600 ha. The habitat for *Pseudosasa usawai* was often found in the smooth slope or less drainage valley but with a great varieties in the elevation. In the level slope area with deep soil, the bamboo can reach over 2 m in height, however, in the ridge area with strong wind and shallow soil, the bamboo height is about 50 cm.

Bamboo shoot of *Pseudosasa usawai* is quite high in economic value and is produced from March to May annually. Therefore, it is harvested by surrounding inhabitants in basic quantity.

(2) Flowering

In Yangminsan national park, a large area was under flowering *Pseudosasa usawai* in 1999 and it lasted till 2000. The bamboo plants died after flowering. Subsequently, seedlings from seeds replaced the dead culms. Till now, the original population has not restored back and consequently, the harvest of bamboo shoots also got affected.

The density of *Pseudosasa usawai* in Yangminshan area is high. After death caused by flowering, a big change in forest micro-environment takes place due to the gap created by canopy loss. The bamboo seedlings in these areas have to compete with other plants that are suitable to grow in the gaps.

(3) Natural Regeneration

Based on the study of seedling growth of *Pseudosasa usawai* in Yangminshan area, bamboo survival curve is close to Type III defined by Deevey (1947) with a high mortality rate in the beginning (Hwang, 2000). The continuous monitoring on the seedling growth showed that the mortality of seedling was decreasing. Referring to Deevey Type III curve, it indicated that the seedling growth in this area passed through a high mortality phase, thinning phase and go to the density stable phase (Hwang, 2001).

3. *Dendrocalamus latiflorus*

(1) The Introducing and History of Planting

Dendrocalamus latiflorus, with pachymorph rhizomes, belongs to *Dendrocalamus* Nees, Genes Gramineae. It was originated from the northern Myanmar. The date of immigration to Taiwan is beyond the records available. It was estimated that about 90,865 ha area under in Taiwan. The culms can be used for a variety of purposes including construction material, bamboo raft, knitting, pulp, etc. Bamboo shoots are quite delicious, and produced in winter and spring seasons. The products of bamboo shoots (*e.g.*, bamboo can) are exported to Japan and USA.

According to the peroxidase isozyme test of 85 culms around Taiwan, the *Dendrocalamus latiflorus* in Taiwan can be classified into four types of isozymic patters. Together with the number of stomata on the leaf, 8 strains are available in Taiwan.

(2) Flowering

The flowering period of *Dendrocalamus latiflorus* is quite long and scattered. According to seed collection experience, the flowering of *Dendrocalamus latiflorus* can last for 6 months. Due to the difference in blossom period, the seed maturity is also scattered which causes difficulty in collecting seeds. Because of the variation in caryopsis maturity, triming of branch or cutting off the whole culms to collect matured seeds, is not possible. It is collected only by hand after checking the seed maturity.

(3) Planting of Seedlings

The purpose of planting seedling of *Dendrocalamus latiflorus* is to develop new generation and preserve old generation gene. Starting from 1984 to 1991, seeds were collected and used to produce seedling plantations at two sites in Linwhachin Experimental Station. The growth survey in 1994 showed that 70 per cent seedlings could develop new culms and the ground diameter of the new culms ranged 3.4-4.5cm with culms height 4.3-5.1m in 1994. Usually, it takes five years to restore back to the original bamboo stands and to produce bamboo shoots again (Lu, 1985; Lu *et al.*, 1997).

4. *Phyllostachys pubescens*

(1) The Introducing and History of Planting

Phyllostachys pubescens. Mazel ex H.de Lehaie, with Leptomorph rhizomes, was originated from Southwest China and was introduced into Taiwan 200 years ago with the planted area of 3,296 ha in 1973. *Phyllostachys pubescens* was planted on a large area in Japan as well.

(2) Flowering of *Phyllostachys pubescens*

The records of flowering in *Phyllostachys pubescens* were observed in China and Japan. In 1970, flowering in *Phyllostachys pubescens* bamboo occurred in Guangsi, Changsi Province, China. Based on the observations in Japan, the blossom cycle for *Phyllostachys pubescens* is 67 years in Kyoto, 48 years in Asakawa and 51 years in Aalanuma. In Taiwan, a small area (0.02 ha) of flowering *Phyllostachys pubescens* was observed in Nanton County (Lu, 1985). Bamboo flowering affects adversely the bamboo management irrespective of the rise of the stand. Consequently, through new generation by sexual propagation, it is desirable to replace the old bamboo from management point of view.

(3) Planting of Seedlings

Since 1985, LU collected seeds from 6 culms. The seed germination rate is high about 69-84 per cent due to the selection of seeds from the well-developed caryopsis. Among them, 70 per cent seedlings were developed into culms. In 1994, the culms ground diameter ranged from 1.3-8.2 cm with the average 4.43 cm. The culms heights are 2.6-12.6 m and 7.2 m in average (Lu and Chen, 1997). While *Phyllostachys pubescens is* Leptomorph bamboos, the seedling is in cluster (Pachymorph). The rhizomes begin to appear at the age of 3.

Conclusions

The flowering in bamboo is quite unique, having a long cycle and with uncertainty. Consequently, researches on bamboo flowering issues are limited. In order to reduce the impact caused by bamboo flowering, the investigations on bamboo flowering mechanisms are essential. For a plant with long period propagation by agamogenesis, the sexual propagation by flowering will benefit the enhancement of the diversity as well as in conservation of the bamboo species.

References

Chen, T. H., Chung, H. S., Wang, D. H., and Wang, J. W. (2010). Growth, biomass and bamboo shoot production of *Pseudosasa usawai*. (In press)

Clark, L. G. (1997). Bamboos: the centrepiece of the grass family. In: Chapman GP (ed) The bamboos. Academic Press, London, pp. 237-248.

Gamble, J. S. (1896). The Bambuseae of British India. Anunals of the Royal Botanic Garden, Culcutta, Vol. VII.

Isagi, Y., Shimada, K., Kushima, H., Tanaka, N., Nagao, A., Ishikawa, T., Onodera, H., and Watanabe, S. (2010). Clonal structure and flowering traits of a bamboo [*Phyllostachys pubescens* (Mazel) Ohwi] stand grown from a simultaneous flowering as revealed by AFLP analysis. *Molecular Ecology* 13, 2017-2021.

Jaksic, F. M., and Lima, M. (2003). Myths and facts on ratadas: bamboo blooms, rainfall peaks and rodent outbreaks in south America. *Austral Ecology* 28, 237-251.

Janzen, D. H. (1976). Why bamboos wait so long to flower. *Annu. Rev. Ecol. Syst.* 7, 347-391.

Lalnunmawia, F., Jha, L. K., and Lalengliana, F. (2005). Preliminary observations on ecological and economical impacts of bamboo flowering in Mizoram (North East India). *Journal of Bamboo and Rattan* 4, 317-322.

Lin, W. C. (1974). Studies on morphology of bamboo flowers. *Taiwan Forestry Research Paper* No. 248, pp. 114.

Lin, W. C. (1976). The classification of subfamily Bambusoideae in Taiwan (continued). *Taiwan Forestry Research Paper* No. 271, pp 75.

Lu, C. M. (1985). Germination and the cultivation of seedlings of giant bamboo (*Dendrocalamus latiflorus* Munro). *Taiwan For Res Inst Note* No. 2.

Lu, C. M., and Chen, C. H. (1997). Silviculture of bamboo seedlings– *Phyllostachys pubescens*. *Taiwan J For Sci* 12, 279-289.

Lu, C. M., Chen, C. H., and Wu, K. W. (1997). Silviculture of bamboo seedlings- *Dendrocalamus latiflorus*. *Taiwan J For Sci* 12, 269-278.

Lu, C. M., and Chen, T. H. (2009). Flowering, fruiting, and recovering regeneration by seedlings of *Melocanna baccifera* in Taiwan. *Bamboo Journal* 26, 56-64.

Makita, A. (1998). The significance of the mode of clonal growth in the life history of Bamboo. *Plant Species Biol* 13, 85-92.

McClure, F. A. (1966). The bamboos, a fresh perspective. Harvard Univ. Press, Cambridge, Massachusetts. (Smithsonian Institute Press, Washington and London)

Shibata, S. (2010). Examination into records and periodicity concerning to *Melocanna baccifera* flowering with 48-year interval. *Jpn J Ecol* 60, 51-62.

Thanchunga, C. (2004). Bamboo flowering in Mizoram, in bamboo. *The Magazine of the American Bamboo Society* 25, 1-3.

Watanabe, M., Ueda, K., Manabe, I., and Akai, T. (1983). Flowering, seeding, germination and flowering periodicity of *Phyllostachys pubescens*. *Journal of Japanese Forest Society* 64, 107-111.

Chapter 7

Ecological Impact Assessment and Predictive Modeling of Bamboo Flowering

S.K. Barik, D. Adhikari and Evanylla Kharlyngdoh

**Department of Botany, North-Eastern Hill University,
Shillong – 793 022, India**

Introduction

Flowering of bamboo is a botanical enigma. Bamboo species flower only once in their lifetime and die subsequently. The factors responsible for it are not clearly understood. Three types of flowering in bamboo have been documented, *viz.* (i) Annual or continuous flowering - those that flower annually or nearly so, *e.g. Arundinaria* spp., *Bambusa lineata, Ochlandra stridula*, etc. (ii) Sporadic or irregular flowering - those that flower sporadically in isolated clumps or in parts of one clump, *e.g. Bambusa balcooa, B. nutans* etc. and (iii) Periodic flowering - those that flower at a time, *e.g. Dendrocalamus hamiltonii, Melocanna baccifera*; both usually flowers sporadically, sometimes gregariously. When all bamboo clumps flower simultaneously in a vast area, it is known as 'gregarious flowering'. Such flowering of bamboo produces large quantities of seeds on which the rodents thrive. Soon after, seed regeneration starts the rats shift to the crop fields for food. This chain is a potential cause for famine. Three types of bamboo death following flowering have been documented *viz.* (i) Flowering does not result in the death of either aerial or underground parts *e.g. Arundinaria, Bambusa atrta* and *Phyllostachys* (ii) Flowering results in complete death of aerial parts only, the rhizome remains alive and plants regenerate *e.g. Arundinaria amabilis, A. simonii* and *Phyllostachys nidularia,* and (iii) Flowering results in complete death of both aerial and underground parts *e.g. Bambusa bamboos, B. tulda, Dendrocalamus hamiltonii, D. strictus* and *Melocanna baccifera.*

The beliefs associated with the phenomenon of bamboo flowering *viz.,* death of the bamboo plant and explosion of rodent population, lacks empirical data. It is generally argued that gregarious flowering of bamboo causes ecological havoc. The bamboo plants die after flowering and it takes a few years before bamboo plants grow up to canopy level again, leaving bare exposed soil, which could be disastrous in mountainous states. The flowering and death of bamboo plants also affect the associated plant species in the surrounding areas. Whether flowering is controlled by environmental or physiological conditions is not clearly established. There is also a strong belief that, albeit its new geographical position and prevailing climatic conditions, the clonal offspring of a bamboo will flower at a definite interval along with the parent plant conforming to its flowering cycle. This assumption, if true, undermines the role of climatic conditions in bamboo flowering. The present study was carried out to test the above assumptions through field surveys and predictive modelling technique in Meghalaya, north-east India. Two bamboo species *viz.,* *Melocanna baccifera* (sympodial) and *Dendrocalamus hamiltonii* (sympodial) were selected for the study (Figures 7.1 and 7.2).

Melocanna baccifera is often referred to by its old name of *Melocanna bambusoides* and it often forms the dominant vegetation in the tropical and subtropical region. *Melocanna baccifera* also known as 'muli' bamboo is a monopodial bamboo with pachymorph (sympodial) rhizome system (Makita, 1998) and grows up to about 20 m height (INBAR, 2004). This species is estimated to represent over a sixth of the country's growing stock of bamboo. The growing stock of this species in Northeast India is worth approximately 50–60 million US dollars if sold as untreated bamboo poles and its financial value as a processed product is estimated at 50 million USD per annum (INBAR, 2004). The culms of *Melocanna baccifera* are thin walled with long internodes and are ideal for splitting, weaving, paper making and construction purposes. The shoots are edible and are used by local communities.

Dendrocalamus hamiltonii is a large sympodial bamboo growing upto 15-20 m height with culm diameter ranging from 12-18 cm. It often forms the dominant vegetation in the humid tropical and some parts of montane subtropical region. In south and southeast Asia, *Dendrocalamus hamiltonii* is one among the most economically important species used for making pulp, paper and rayon, building construction, fencing, ceiling, walling native huts, scaffolding, basket-making, fuel, handicrafts, and floats for timber-rafts. It is also used in the development of farm-oriented cottage industry in almost every village. The young shoots are edible and culms are also used as medicine for curing fever and food poisoning (Wang *et al.,* 2002). Besides its economic values, *Dendrocalamus hamiltonii* stands can also hold equivalent or higher levels of plant-carbon (71 t C/ha) compared to *Eucalyptus* (30-72 t C/ha) stands (INBAR, 2009). Ecological impact of flowering of their two important bamboo species of Meghalaya is not well-understood. Therefore, the prsent study assessed the impact of flowering on species compositin of the forest. An attempt was also made to assess if the flowering event is a function of environmental factors or not through applying Ecological Niche Modelling tool.

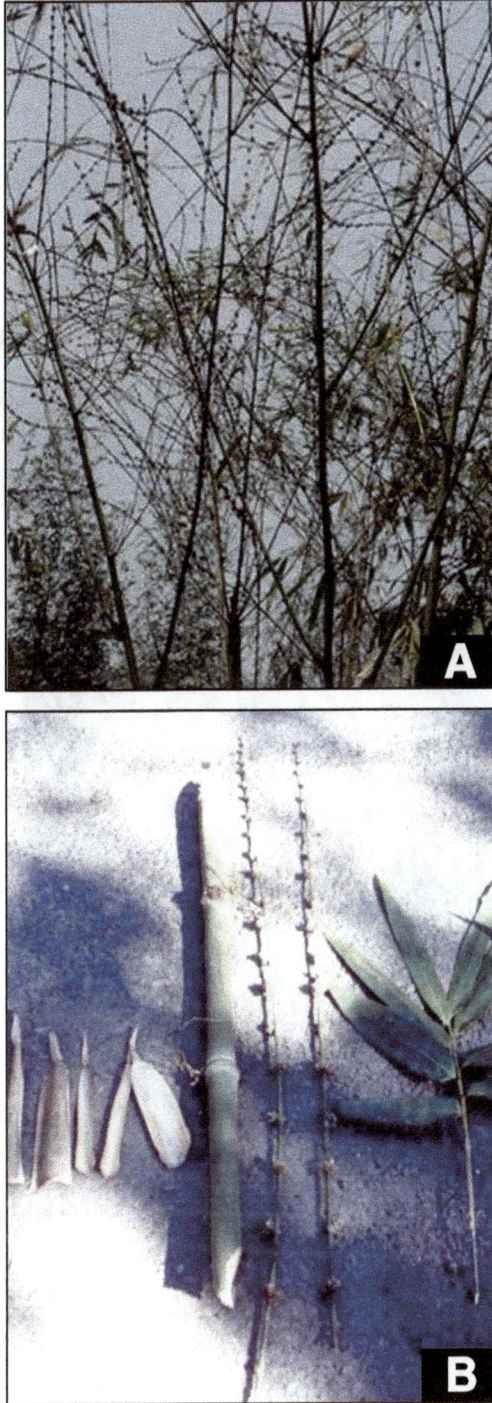

Figure 7.1: *Dendrocalamus hamiltonii*
(A) Flowering clump, (B) Culm sheath, culm, inflorescence and leaves.

Figure 7.2: *Melocanna baccifera* - (A) Inflorescence, and (B) Fruits.

Methods

Assessment of Ecological Impact

Ecological impact of flowering of two bamboo species *viz., Dendrocalamus hamiltonii* (sympodial) and *Melocanna baccifera* (monopodial) in Meghalaya, north-east India was evaluated. The changes in species composition and pattern of recovery following bamboo flowering were monitored over a period of three years beginning from 2007. In order to monitor the species composition, random sampling of vegetation was done in the flowering sites of Garo Hills along Tura-Baghmara Highway by laying quadrats of 10 x 10 m size for trees, 5 x 5 m for shrubs and 1 x 1 m for herbs. The plants of different species in each quadrat was identified and counted. Based on the quadrat data, frequency, density and cover (basal area) of each species were calculated. The importance value index (IVI) for different tree species were determined by summing up the Relative Density, Relative Frequency and Relative Cover values. The Relative Density and Relative Frequency values were used to calculate the IVI of shrubs and herbs.

Predictive Modelling of Bamboo Flowering

Geographic coordinates of the spatial incidences of flowering for *Dendrocalamus hamiltonii* and *Melocanna baccifera* in Meghalaya for the years 2007 - 2010 were recorded through field visits.

MODIS images (MOD13Q1) with a spatial resolution of 500 m (obtained from Oak Ridge National Laboratory Distributed Active Archive Centre (http://daac.ornl.gov/) were used as predictors of bamboo flowering events. These layers correspond to the years 2007 to 2010, and characterize the aggregates of Enhanced Vegetation Index (EVI) at 16 days interval.

The geographical coordinates of flowering localities were divided into two parts *i.e.* 2007-2008 and 2009-2010, and the niche models were built independently from the two data partitions. For building different subset models of the entire area of occurrence, an algorithm threshold of 0.01 was selected with 1000 iterations as an upper limit for each run. 50 per cent of the total occurrence points were used to generate the rule and the remaining 50 per cent was used for intrinsic testing for model optimization and refinement. Due to the stochastic nature of GARP in giving different outputs at different runs the model was set to perform 100 tasks. For selecting the 10 best subset models, the inbuilt best subset selection button in Desktop GARP was checked with threshold level of 0 per cent extrinsic hard omission and 50 per cent commission. All the 10 best subset models were summed up using the calcgridsinview.ave script in ArcView 3.3 to create a composite distributional representation of the ecological niche of *D. hamiltonii*. The independent validation of the model was done by overlaying the occurrence points with the model projections. After model calibration, the final predictions were made by projecting the rules to Meghalaya. In order to make the final model projections, GARP was set to perform 100 tasks with a threshold level of 0 per cent extrinsic hard omission and 50 per cent commission.

To assess the predictive ability of the ecological niche model in a temporal range, the predictive maps were subjected to independent validation test. This was done by overlaying the independent data points for outbreak locations on the final model projection.

Results and Discussion

Distribution of *Melocanna baccifera* and *Dendrocalamus hamiltonii*

Melocanna baccifera is found at an elevational range of 100-1200 m a.s.l. in Meghalaya and other north-eastern states of India. Worldwide, it is distributed in Bangladesh, China, Florida, Gautemala, Jamaica, Myanmar, Thailand, and Singapore. In India, it is found in the states of Arunachal Pradesh, Assam, Dehradun (FRI), Manipur, Meghalaya, Mizoram, Nagaland, Sikkim, Tripura, West Bengal. In Meghalaya, it is distributed in East Khasi Hills: Mawmluh; Jaintia Hills; East Garo Hills: Bajengdoba, Jengjal, Mendal, Samphalgiri, Sugiri; Ri-Bhoi: Nongpoh; West Garo Hills: Garobada, Modinagar, Tura; West Khasi Hills: Nongshram, Shalang; South Garo Hills: Dalu, Gasuapara.

Dendrocalamus hamiltonii occurs at an elevational range of 0-1700 m a.s.l. in north-eastern India. Worldwide, it is distributed in Bangladesh, Bhutan, China, Laos, Nepal, Upper Myanmar, and Vietnam. In India, it is found in the states of Andhra Pradesh, Arunachal Pradesh, Assam, Bihar, Dehradun, Manipur, Meghalaya, Mizoram, Nagaland, Sikkim, Tripura, Uttar Pradesh, West Bengal. In Meghalaya, it is distributed in East Khasi Hills: Garampani, Nongshken, Pynursla, Tangmang, Shella; Jaintia Hills: Dawki, Jowai, Ladrymbai, Pasadwar, Ratachera, Sonapur, Syndai, Umkiang; East Garo Hills: Duragiri, Rongrenggiri; Ri-Bhoi: Arphewmer, Erpakon, Lasharai, Nongkhyllem, Nongpoh, Patharkhmah, Umbang, Umling, Umran, Umroi, Umsning; South Garo Hills: Baghmara, Bajengdoba, Rangung, Songsak, Williamnagar; West Garo Hills: Salmanpara, Manchor; West Khasi Hills: Domiasiat, Jirang, Mairang, Mawdoh, Mawthylliang, Nongkhlaw, Nonglang, Nongpyndeng, Nongshiliang, Nongshram, Nongstoin, Pambriew, Porla, Pynnohumiong, Rwiang river, Seinduli, Shalang, Sonapahar.

Flowering of *Melocanna baccifera* and *Dendrocalamus hamiltonii* in the Indian Region

Melocanna baccifera

The first recorded flowering of *Melocanna baccifera* was in 1863 from Cachar, Assam. Various periods of vegetative growth prior to flowering have been noted in different locations and there is a wealth of reports that give a good, reliable picture of the flowering cycle of this species. Both sporadic and gregarious flowering types have been observed in the species. The flowering period of this species has been reported at the beginning of dry season from September to January and the flowering cycle of the species ranges from 30 to 60 years (Table 7.1). According to INBAR (2004) flowering factsheets, three main flowering cycles have been reported: 30-35 years, approximately 45 years, and 60 years. Mizoram, Assam and parts of Mizoram and Meghalaya (parts of the Garo hills) and the north of Tripura are on a short flowering cycle of 30-35 years, whilst those in the remainder of Mizoram, Chittagong hill tracts and the remainder of the Garo hills are on longer flowering cycles of between 40-50 years. The regular once-every-48-year flowering in Mizoram is well recorded. Muli bamboo flowering at FRI Dehra Dun in 1992 was reported to have come from a population that flowered in 1960, which was established through seedling introduction in 1912. This gives one cycle of 48 years. Flowering of successive generations in Haflong in Assam is reported at 40- 44 years (flowering in 1910-12 and 1952-6) and 32-36 years (flowering in 1988). There are also some reports of flowering after 7-10, 19-21, 25-27 years.

Table 7.1: Flowering History of *Melocanna baccifera* in the Indian Region.

State	Flowering Years
Assam	1863, 1866, 1892, 1893, 1900-1902, 1910-1912, 1933, 1960 (Cachar- reported by Chatterjee, 1960; Vaid, 1972), 1967 (Cachar- reported by Nath, 1968), 1892, 1952-1956 (Haflong), 1998, 2006, 2008 (Bajali- reported by Sarma *et al.*, 2010)
Dehradun	1992 (FRI- reported by Sharma, 1992)
Manipur	1952, 1967 (reported by Nath, 1968), 2003-2008 (Tamenglong)
Meghalaya	1815, 1863, 1911, 1959 (reported by Pathak and Kumar, 2000) 2004-2010 (Garo Hills), 2006, 2007 (East Khasi Hills)
Mizoram	1815, 1863-1864, 1911, 1958-1959 (reported by Pathak and Kumar, 2000), 1960, 2001-2005 (reported by Government of Mizoram), 2006-2010
Nagaland	1960, 2007, 2008
Pune	1993
Sikkim	1815, 1863, 1911, 1959 (reported by Pathak and Kumar, 2000)
Tripura	1992-2002, 2003-2009, 2010
West Bengal	1960 (Siliguri; reported by Chatterjee, 1960)
Flowering period	September-January
Flowering types	Sporadic and gregarious
Flowering intervals	Short interval: 30-35; Long intervals: 40-50 or upto 60 years 30 years (Gamble, 1896); 30-35 years (Kurz 1876); 45 years (Troup 1921); 40-45 years (Seethalakshmi and Kumar, 1998). 7-10, 19-21, 25-27, 30-35, 45-50, 60 year intervals (INBAR website)

Dendrocalamus hamiltonii

Records of flowering in *Dendrocalamus hamiltonii* dates back to 1894 from Sikkim and Dehradun. Subsequently, numerous instances of flowering have been reported from Assam, Punjab and Meghalaya. The species flowers sporadically and sometimes gregariously with intervals of 30-40 years and has been recorded to flower mostly from October-May (Table 7.2). Soon after profuse flowering and seeding, the entire clump dies. About 153 to 210 seedlings per square meter were observed and one and two year old seedlings recorded a height of 12 and 18 cm respectively (Seethalakshmi and Kumar, 1998).

Characteristics of *Melocanna baccifera* before and during Flowering Period

Two years prior to flowering, the production of new shoots was reduced from 19 to 8 shoots/m². Further, during the flowering period, the number of new shoots

Table 7.2: Flowering History of *Dendrocalamus hamiltonii* in the Indian Region.

State	Flowering Years
Arunachal Pradesh	1994-2005; 2003-2005 (Roin, Dibang valley- reported by Government of Arunachal Pradesh),
Assam	1905 (Lakhimpur; reported by Cavendish, 1905), 1912, 1956, 1981-82, 1997-1998 (Golaghat- reported by Pathak), 2001-2007
Dehradun	1894, 1992
Manipur	2001-2005
Meghalaya	1921, 1991, 2000 (reported by Pathak and Kumar, 2000), 2005, 2008
Mizoram	1921, 1991, 2000 (reported by Pathak and Kumar, 2000), 2001-2002, 2004 (reported by Government of Mizoram) 2001-2010
Nagaland	2003-2004 (reported by Government of Nagaland)
Punjab	1992
Sikkim	1894, 1921, 1991, 2000 (reported by Pathak and Kumar, 2000)
Tripura	2010 (reported by INBAR, 2010)
West Bengal	1976 (reported by Conservator of Forests, 1976)
Flowering period	October-May
Flowering types	Sporadic and sometimes gregarious
Flowering intervals	30-40 years

emerged was nil. This finding was also supported by Banik (1989) who reported that shoot production in observed clumps in Bangladesh in the 1980s gradually reduced in the three years prior to flowering from 17 new culms to 8 per clump, and in the year of flowering no new culms emerged at all.

During the flowering period, every part of the plant produced flowers, including the rhizomes. And shortly after flowering, all the branches of the culms were devoid of leaves and every bud starts producing reproductive organs. A large fruit was developed after flowering period. The fruits of culms that flower during the month of November generally ripe and mature in about six months time during April-May. Whereas, fruits of florets flowering in the earlier part of the season mature more rapidly than those flowered later. The fruits were attached to the parent culms by their thicker ends, with the pointed end often hanging downwards usually for few months.

The fruits were fleshy, large, ovoid or globose with a long extended beak, and resembled a green guava. The fruit production per square meter was 120 ±15, and the average fruit weight was 96 ±6.0 g per fruit. The average fruit length and diameter was 7.0 ±1.0 cm and 5.0 ±0.5 cm, respectively (Table 7.3). Similar results were also reported by Chatterjee (1960) who mentioned that the fruits of *Melocanna baccifera* were between 6-12 cm long and up to 8 cm wide, and weigh up to 275g when fresh, although most were in the range of 75-150g.

Table 7.3: Characteristics of *Melocanna baccifera* before, during and after Flowering.

YEAR BEFORE FLOWERING	
Study area (ha)	2.0
Number of live shoots/m²	8 ±3.0
Number of live culms/m²	17 ±4.0
Total culms in 2 ha	3,42,500
Avg. culm height (m)	8.0 ±2.0
Avg. culm diameter (cm)	2.0 ±0.5
Leaf shedding period of mature culm	October-March
Leaf flushing period of mature culm	April
DURING FLOWERING/FRUITING	
Flowering period	September, 2006
Flower colour	Reddish pink
Fruiting period	November, 2006
Number of fruits/culm	8 ±4
Avg. fruit weight (g)	96 ±6.0
Avg. fruit diameter (cm)	5.0 ±0.5
Avg. fruit length (cm)	7.0 ±1.0
Fruit production/m²	120 ±15
Total fruit production in 2 ha	23,97,500
REGENERATION	
Seedling emergence	April-May, 2007
Number of seedlings/m²	11 ±3
Total seedlings in 2 ha	2,20,000
Avg. height of a 3 months old seedlings (cm)	30.0 ±5.0
Avg. height of a 3 years old juvenile culm (m)	3.0 ±0.6

The fruits of *Melocanna baccifera* remained viabile up to 40-60 days under normal conditions and germinated within few weeks of release from the mother plant. Both radicle and plumule emerged simultaneously. The number of seedlings per square meter was 11±3. The young seedlings reached a height of 30 ±5.0 cm in three months period, whereas that of a 3 years old juvenile culm reached up to 3.0 ±0.6 m long (Table 7.3). Vivipary was also observed in a few culms.

The time period from flowering and seeding to death for an individual plant completed in a maximum of one year. After the flowering and fruiting period was over, no parent culms or rhizomes survived. This finding was also supported by INBAR (2004) who stated that there are no reports of *Melocanna baccifera* plants surviving flowering.

Ecological Impact of Bamboo Flowering

The changes in species composition and pattern of recovery following bamboo flowering were monitored over a period of three years beginning 2006. The flowering of *Dendrocalamus hamiltonii* which was sporadic in nature did not impact species composition in the forest stand substantially. On the other hand, *Melocanna baccifera* which experienced gregarious flowering in 2007 significantly altered the species composition of the bamboo forest (Tables 7.4–7.6). Due to gregarious flowering and subsequent death of both the aerial parent culms and underground rhizomes, gaps of different sizes and shapes were created in the forest canopy. This further altered the forest microclimatic conditions significantly affecting community dynamics (Orians, 1982; Nunez–Farfan and Dirzo, 1988; McCarthy and Facelli, 1990). Although bamboos compete effectively with other tree and shrub species, their natural death may prevent them from being dominant for atleast a few years, thus allowing other light demanding species to coexist (Taylor and Zisheng, 1992; Marod *et al.,* 1999; Abe *et al.,* 2002; Taylor *et al.,* 2004; Holz and Veblen, 2006).

Table 7.4: Tree Species Sssociated with *Melocanna baccifera* before and after Flowering Period.

Species	Family	IVI	
		Before Flowering	After Flowering
Bridelia monoica	Euphorbiaceae	9.41	9.41
Cinnamomum tamala	Lauraceae	17.04	17.04
Cinnamomum zeylanicum	Lauraceae	22.36	22.36
Eurya acuminata	Theaceae	9.41	9.41
Ficus spp.	Moraceae	17.36	17.36
Garcinia lancifolia	Guttiferae	19.40	19.40
Mallotus spp.	Euphorbiaceae	22.63	22.63
Musa spp.	Musaceae	26.83	26.83
Rhynchotechum ellipticum	Gesneriaceae	16.77	16.77
Ricinus communis	Euphorbiaceae	16.18	16.18
Schima wallichii	Theaceae	14.40	14.40

Table 7.5: Shrub Species Associated with *Melocanna baccifera* before and after Flowering Period.

Species	Family	IVI	
		Before Flowering	After Flowering
Camellia caduca	Theaceae	5.87	5.52
Chromolaena odorata	Asteraceae	10.47	10.59
Clerodendrum colebrookianum	Verbenaceae	6.52	6.45
Clerodendrum viscosum	Verbenaceae	4.42	4.53
Crotalaria juncea	Fabaceae	7.97	8.13
Derris thyrsiflora	Fabaceae	5.51	5.47
Desmodium spp.	Fabaceae	4.42	4.53

Contd...

Table 7.5–*Contd...*

Species	Family	IVI	
		Before Flowering	After Flowering
Dicranopteris linearis	Pteridaceae	13.62	14.21
Ficus hispida	Moraceae	3.99	4.07
Leea indica	Leeaceae	7.43	7.55
Lonicera macrantha	Caprifoliaceae	4.78	4.82
Maesa indica	Myrsinaceae	7.97	8.13
Maesa ramentacea	Myrsinaceae	6.30	6.45
Melocanna baccifera	Poaceae	53.44	51.52
Mimosa pudica	Fabaceae	7.28	7.27
Mussaenda macrophylla	Rubiaceae	6.09	6.22
Phyllanthus spp.	Euphorbiaceae	6.30	6.45
Phyllanthus glaucus	Euphorbiaceae	7.54	7.67
Saccharum spontaneum	Poaceae	7.57	7.57
Thysanolaena maxima	Poaceae	10.43	10.89
Urena lobata	Malvaceae	6.09	6.22
Xeromphis spinosa	Rubiaceae	5.98	5.75

Table 7.6: Herbs Species Associated with *Melocanna baccifera* before and after Flowering Period.

Species	Family	IVI	
		Before Flowering	After Flowering
Aconogonum molle	Polygonaceae	7.11	6.15
Ageratum conyzoides	Asteraceae	8.77	7.93
Arundinella nepalensis	Poaceae	13.10	10.27
Centella asiatica	Apiaceae	8.66	9.80
Cissampellos pareira	Menispermaceae	2.83	2.15
Commelina nudiflora	Commelinaceae	3.64	2.14
Conyza stricta	Asteraceae	2.57	2.90
Cyperus roduntus	Cyperaceae	12.03	10.76
Desmodium gangeticum	Fabaceae	2.94	3.05
Dioscorea alata	Dioscoreaceae	2.56	2.91
Dioscorea heterophylla	Dioscoreaceae	2.89	2.08
Eriocaulon cristatum	Poaceae	6.42	6.62
Eupatorium adenophorum	Asteraceae	4.54	4.77
Hedychium gardnerium	Zingiberaceae	3.58	4.08

Contd...

Table 7.6–*Contd...*

Species	Family	IVI Before Flowering	IVI After Flowering
Hedyotis scandens	Rubiaceae	2.08	2.28
Hypochaeris radicata	Asteraceae	2.94	3.52
Imperata cylindrica	Poaceae	10.91	10.62
Jasminum spp.	Oleaceae	2.03	2.21
Lygodium flexuosum	Schizaceae	3.10	3.46
Merremia umbellta	Convolvulaceae	2.62	2.98
Merremia vitifolia	Convolvulaceae	2.51	2.28
Mikania micrantha	Asteraceae	7.75	8.78
Molineria palmifolia	Poaceae	5.13	5.11
Mucuna bracteata	Fabaceae	3.63	4.15
Osbeckia stellata	Melastomaceae	3.05	3.18
Paederia scandens	Rubiaceae	4.01	3.95
Paspalum dilatatum	Poaceae	10.64	10.34
Passiflora nepalensis	Passifloraceae	2.73	2.98
Phrynium pubinerve	Marantaceae	4.22	3.80
Pilea symmeria	Urticaceae	6.47	6.35
Polygala arvensis	Polygalaceae	2.94	3.25
Polygonum chinensis	Polygonaceae	5.13	5.11
Pteridium acquilinum	Pteridiaceae	2.94	2.21
Renanthera imschootiana	Orchidaceae	3.85	4.37
Scleria levis	Cyperaceae	3.26	3.66
Selaginella spp.	Selaginellaceae	6.52	8.00
Smilax glabra	Smilacaceae	2.67	2.77
Smilax perfoliata	Smilacaceae	2.83	1.94
Solena amplexicaulis	Cucurbitaceae	3.53	3.18
Stephania glandulifera	Menispermaceae	2.78	4.57
Tetrastigma bracteolatum	Vitaceae	4.28	4.98
Thladiantha calcarata	Cucurbitaceae	3.79	4.36

In *Melocanna baccifera* stands, the IVI of herbaceous species varied significantly before and after flowering, whereas the IVI of shrubs species did not vary much. The IVI of the tree species however, were the same both before and after flowering period. Among the associated plant species, *Mikania micrantha, Thysanolaena maxima* and *Imperata cylindrica* dominated the *Melocanna baccifera* stands till third year. The seedlings of a few pioneer broad-leaved species such as *Callicarpa arborea, Eurya acuminata* and *Schima wallichii* accelerated their growth following the flowering of

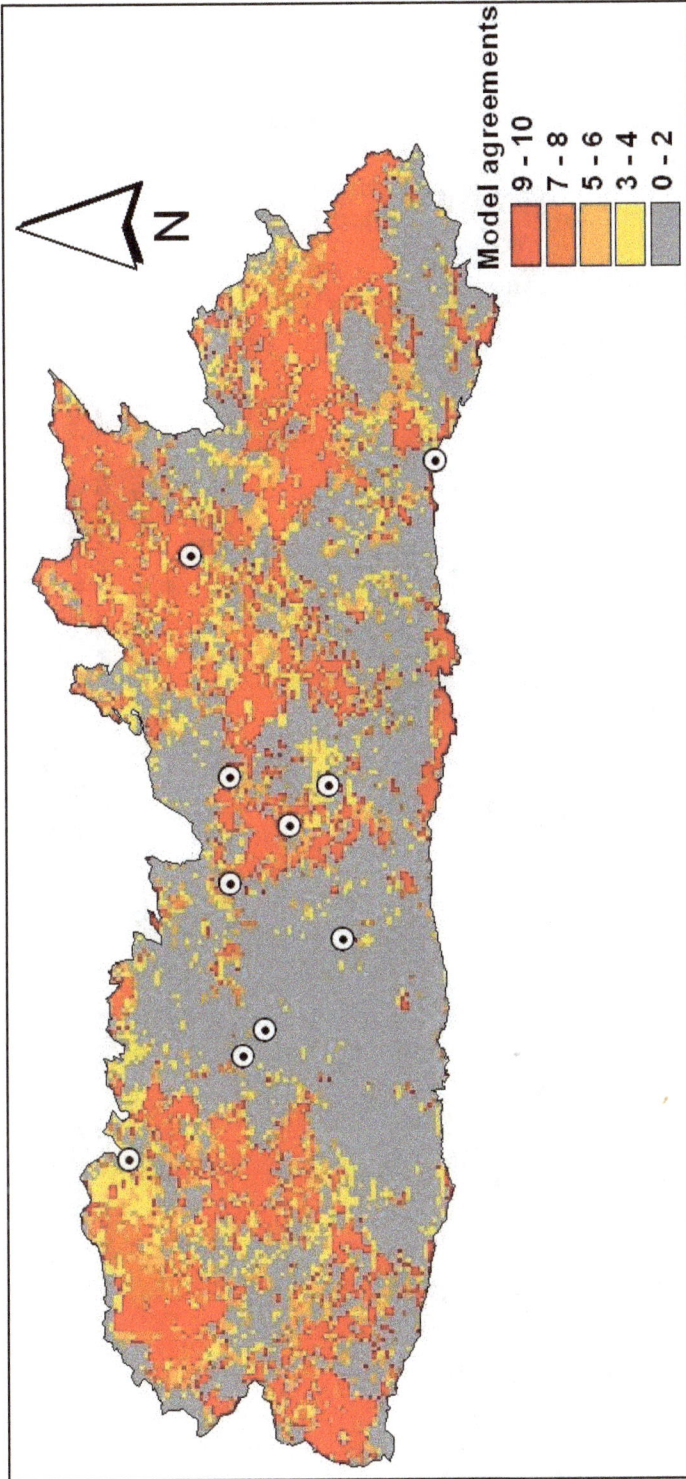

Figure 7.3: Flowering of *D. hamiltonii* in 2007-08 Predicting 2009-10.

Circle with dots represent the sporadic flowering events recorded in the year 2009-2010. The color ramp represents the model agreements where the red color represents the areas where 9-10 models predict environmental suitability for flowering, and the grey color represent the areas where 0-2 models predict environmental suitability.

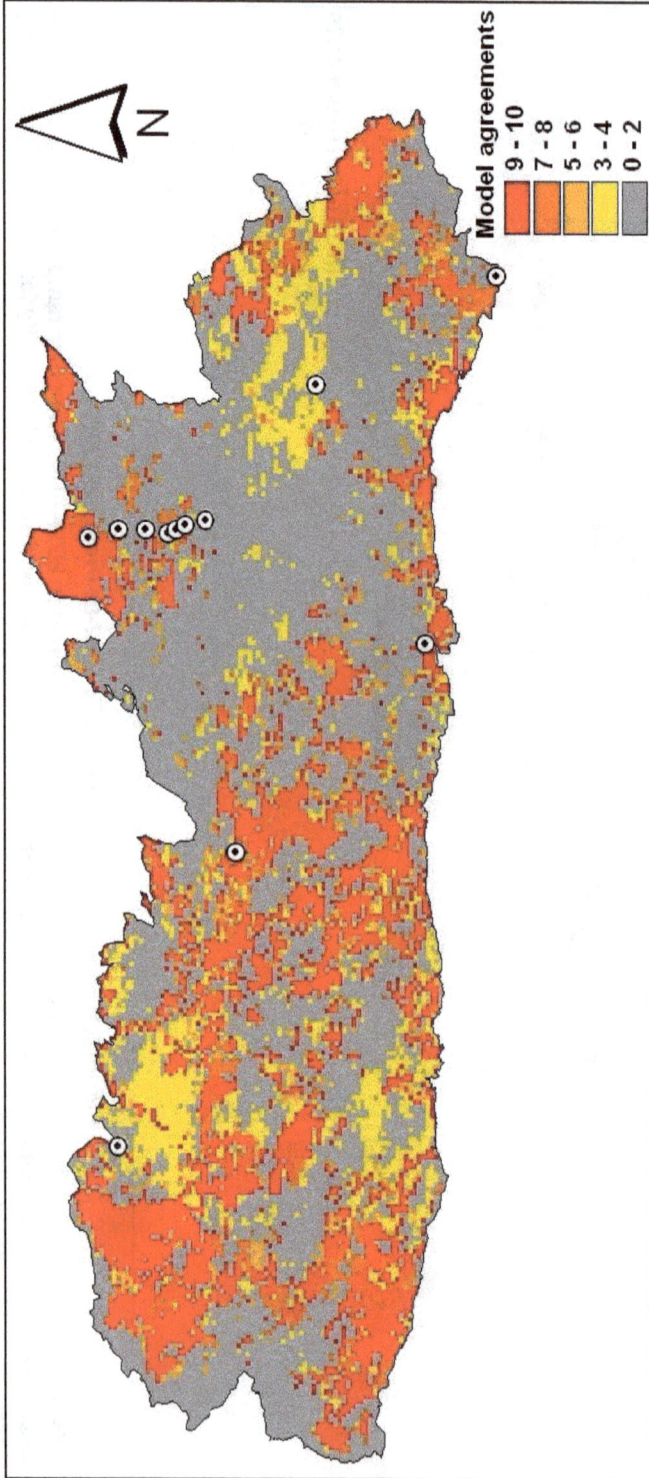

Figure 7.4: Flowering of *D. hamiltonii* in 2009-10 Predicting 2007-08.

Circle with dots represent the sporadic flowering events recorded in the year 2007-08. The color ramp represents the model agreements where the red color represents the areas where 9-10 models predict environmental suitability for flowering and the grey color represent the areas where 0-2 models predict environmental suitability.

bamboo, a concept similar to advanced regeneration mechanism of shade-tolerant tree species in primary forests following small scale disturbance.

Predicting Flowering Events in *Dendrocalamus hamiltonii*

The objective of predictive modelling was to test whether bamboo flowering occurs under a predictable environmental setting. The study used time series flowering data for *Dendrocalamus hamiltonii* for 2007-2008 and 2009-2010 for model building and validation and enhanced vegetation index (EVI) as predictors of flowering. Models built using flowering data points for the year 2007-2008 predicted 6 out of 10 flowering localities of the year 2009-2010, while models built using the flowering data points for the year 2009-2010 predicted 8 out of 13 localities of the year 2007-2008 (Figures 7.3 and 7.4 and Table 7.7). Reciprocal projection tests predicted >60 per cent of the flowering events indicating the importance of climatic factors in bamboo flowering. However, the models are required to make more robust by confirming the parental origin, enhancing the geographical area beyond Meghalaya and including more data sets from diverse agro-climatic conditions. The initial results of this study indicate that there exists a relationship between bamboo flowering and prevailing environmental conditions.

Table 7.7: Model Results for Reciprocal Prediction Tests for *D. hamiltonii.*

Year	Model Calibration	No. of Data Points Used for Independent Model Validation	No. of Events Correctly Predicted
2007-08 predicts 2009 -10	13	10	6
2009-10 predicts 2007 -08	10	13	8

References

Abe, M., Izaki, J., Miguchi, H., Masaki, T., Makita, A. and Nakashizuka, T. (2002). The effects of *Sasa* and canopy gap formation on tree regeneration in an old beech forest. *Journal of Vegetation Science* 13: 565–574.

Banik, R.L. (1989). Recent flowering of muli bamboo (*Melocanna baccifera*) in Bangladesh: An alarming situation for the bamboo resource. *Bano Biggyan Patrika* 18 (1-2): 65-68.

Cavendish, F.H. (1905(. A flowering of *Dendrocalamus hamiltonii* in Assam. Indian Forester 31 (8): 479.

Chatterjee, D. (1960). Bamboo fruits. *Journal of the Bombay Natural History Society* 57 (2): 451-453.

Conservator of Forests, West Bengal (1976). Gregarious flowering of bamboos-*Dendrocalamus hamiltonii. Indian Forester* 102 (11): 83.

Gamble, J.S. (1896). *The Bambuseae of British India. Annals of Royal Botanic Garden,* Calcutta, volume 7, 133 pp.

Government of Arunachal Pradesh

Government of Mizoram

Holz, C.A. and Veblen, T.T. (2006). Tree regeneration responses to *Chusquea* montana bamboo dieback in a subalpine *Nothofagus* forest in the southern Andes. *Journal of Vegetation Science* 17: 19–28.

INBAR (2004). *Melocanna baccifera* flowering factsheets. International Network for Bamboo and Rattan, Beijing, China, pp 1-4.

INBAR. International Network for Bamboo and Rattan Website (available at www.inbar.int).

INBAR (2009). Capturing carbon with bamboo: fast and effective in managed stands. INBAR environment factsheet No. 3. Cop 15, Copenhagen, December 7–18, pp 1-2. International Network for Bamboo and Rattan, Beijing, China.

INBAR (2010). Annual report: In partnership for a better world. International Network for Bamboo and Rattan, Beijing, China, pp 1-32.

Kurz, S. (1876). Bamboo and its use. *Indian Forester* 1: 2 19-69.

Makita, A. 1998. The significance of the mode of clonal growth in the life history of bamboos. *Plant Species Biology* 13: 85-92.

Marod, D., Kutintara, U., Tanaka, H. and Nakashizuka, T. (1999). Structural dynamics of a natural mixed deciduous forest in western Thailand. *Journal of Vegetation Science* 10: 777–786.

McCarthy, B.C. and Facelli, J.M. (1990). Micro disturbances in old fields and forests: implications for woody seedling establishment. *Oikos* 58: 55–60.

Nath, G.M. (1968). Flowering of muli bamboos (*Melocanna bambusoides*). *Indian Forester* *94: 346.*

Nunez–Farfan, J. and Dirzo, R. (1988). Within gap heterogeneity and seedling performance in a Mexican tropical forest. *Oikos* 51: 274–284.

Orians, G.H. (1982). The influence of tree falls in tropical forests on tree species richness. *Tropical Ecology* 23: 255–279.

Pathak, K.C. Gregarious bamboo flowering in northeastern region and some management considerations. Expert consultation on bamboo flowering, Rain Forest Research Institute, Jorhat.

Pathak, K.A. and Kumar, D.K. (2000). Bamboo flowering and rodent outbreak in north eastern hill region of India, *Indian Journal of Hill Farming* 13: 1–7.

Sarma, H., Sarma, A.H., Sarma, A. and Borah, S. (2010). A case of gregarious flowering in bamboo, dominated lowland forest of Assam, India: phenology, regeneration, impact on rural economy and conservation. *Journal of Forestry Research* 21(4): 409"414.

Seethalakshmi, K.K. and Kumar, M.S.M. (1998). *Bamboos of India: a compendium.* Kerala Forest Research Institute, Peechi and International Network for Bamboo and Rattan, Beijing, People's Republic of China, 342 pp.

Sharma, A. (1992). Muli bamboo flowers in FRI, Dehradun. *Indian Forester* 118 (11): 862-864.

Taylor, A.H. and Zisheng, Q. (1992). Tree regeneration after bamboo die back in Chinese *Abies–Betula* forests. *Journal of Vegetation Science* 3: 253–260.

Taylor, A.H., Jinyan, H. and Shiqiang, Z. (2004). Canopy tree development and undergrowth bamboo dynamics in old growth *Abies–Betula* forests in south-western China: a 12 year study. *Forest Ecology and Management* 200: 347–360.

Troup, R.S. (1921). *The silviculture of Indian trees, volume III, Lauraceae to Coniferae.* The Clarendon Press, Oxford.

Wang, K., Hong, L.T. and Rao, V.R. (2002). Diversity and indigenous utilization of bamboo in Xishuangbanna, Yunnan Province, Southwest China. *Journal of Bamboo and Rattan* 1 (3): 263–273.

Chapter 8

Bamboo Flowering Related Upsurge in Rodents and their Management in North-Eastern India

N.S. Azad Thakur, K. Saikia and D. Kumar

Division of Entomology, ICAR Research Complex for NEH Region, Umiam, Meghalaya, India

Introduction

Bamboos occur in great abundance and variety in the South and Southeast Asian countries. India has the distinction of being the second highest bamboo producing country after China and more then 55 per cent of the crop is produced in Northeast. Bamboo spreads over an area of 10.03 million hactares, which constitutes 12.8 per cent of the total forest cover of India. The total bamboo cover in the north-eastern hill (NEH) region has been reported to be 38,197 km² (Annonymous, 1999). Among seven sisters of NEH Region, Mizoram occupies the largest forest area under bamboo (30.8 per cent), followed by Tripura (27.13 per cent) and Meghalaya (26 per cent). There are 16 genera and 58 species of bamboo in the region (Bahadur and Jain, 1981) and about 26 different types of bamboo are reported to grow in Mizoram. *Melocanna baccifera* dominates other species and constitutes approximately 95 per cent of the total growing stock in Mizoram. In the region, bamboo plays an important role in rural and socio-economic development of the indigenous communities. Tribal communities of this region use this potential resource for food, shelter, furniture, handicraft, medicines and various ethonoreligious purposes. Owing to multifarious nature of bamboo resource, it is considered a minor forest produce of high value. Gregarious and mass flowering of few dominant commercial bamboo species are the natural threats for the drastic change in their population structure, productivity and young shoots. Gregarious flowering of *Melocanna bacifera* is believed to take place after 48-50 years resulting in famine. As the flowering of this species was reported in 1958-59, thus the same and also of *Bambusa cacharensis* was expected to occur between 2005-2007 (Singh *et al.,* 2003).

This spectre of gloom is haunting the NE region once again and the natural phenomenon is occurring in different parts of the North-East. Gregarious flowering of Muli bamboo is spread over an area of 18,000 sq kms in India's north-eastern states of Mizoram – the epicentre, Tripura, Manipur, Arunachal Pradesh, Meghalaya and (parts of) Assam.

Bamboo Flowering

The mechanism for the timing of flowering and dying is a phenomenon not yet fully understood. Some bamboos flower only infrequently, others only rarely and some never flower in human memory. Depending upon the species, the flowers appear at intervals of several years if not decades or centuries. It is one of nature's baffling mysteries. Since bamboo is anemophilous and wind pollinated, it must have many flowers at anthesis at the same time for successful spread of the pollen.

As observed, bamboo flowering starts in October-December, immediately after the rainy season (September-October). Initially there are many young inflorescences. Within a few weeks the whole clumps get transformed into huge inflorescences. There is wind pollination and seed-set in January-February and the seed shed starts in February-March. By March end, there is a thin layer of seeds on the forest floor below the bamboo stands. In the following (March-May) months during summer, there is an increase in seed-shed. By the end of summer there are enough seeds on the forest floor. With the onset of rains, bamboo seeds germinate in a few days and the layer of bamboo seeds vanishes and a lush green carpet of bamboo seedlings can be seen. It is reported that a 40 square yard clump of the Indian *Dendrocalamus strictus* produces 145.3 kg of seeds and there are 800–1000 seeds to about 30 g.

(i) Historical Background

Two types of periodical bamboo flowerings *viz., Mautam* and *Thingtam* are reported from Mizoram. Mautam is associated with "*Mao*" (*Melocanna baccifera*), which flowers in a 48 ± 2 years cycle and *Thingtam* is associated with flowering of "*Rawnal*" *(Dendrocalamus longispathus)* and "*Rawthing*" *(Bambusa tulda)* species. The gap between "*Mautam*" to "*Thingtam*" is 18 ± 2 years and between "*Thingtam*" to "*Mautam*" is 28-30 years approximately (Table 8.1).

Similar mass flowering was recorded in Garo hills, Meghalaya (1920-21), Nagaland (1929-30), Manipur (1954-55), North Cachar district of Assam (1958) and East Kameng district of Arunachal Pradesh (1991). Massive flowering in 1998 winters and a corresponding increase in rat population was reported to cause severe damage to crops in Tamenglong, Churachandpur and Ukhrul districts of Manipur.

Table 8.1: Periodic Bamboo Flowering in Mizoram.

Bamboo in Flower	Period	Type
Bambusa tulda Dendrocalamus longispathus	1880-1884	Thingtam
Melocanna baccifera	1910-1912	Mautam
Bambusa tulda Dendrocalamus longispathus	1924-1929	Thingtam
Melocanna baccifera	1958-1959	Mautam
Bambusa tulda Dendrocalamus longispathus	1976-1977	Thingtam
Melocanna baccifera	2005-2007	Mautam

(ii) Present Scenario

Two types of bamboo *viz., M. baccifera* and *D. hamiltonii* started to bloom in 2004 in Garo hills of Meghalaya, Kolasib district of Mizoram; three kinds of bamboo — *Jati, Muli* and *Pessa* in Langtin and Hatikali areas in Assam and four varieties *viz.,* Talam (*Schizostachyum fuchsianum*), Tadar (*Schizostachyum* sp), Aaye and Taffo (*Phyllostachys* sp.) flowered gregariously, while sporadic flowering was observed on Hubbi, Hegang and Haajo in Arunachal Pradesh.

Rodent Species Diversity during Bamboo Flowering

Surveys conducted in different areas of Mizoram and Meghalaya showed that there was sporadic to mass bamboo flowering of *Melocanna baccifera* in both the states but no upsurge of rodents was noticed. During the survey, rats collected from bamboo flowering areas were identified from ZSI, Kolkota (Table 8.2). It was observed that the rodent activities increased in paddy fields/jhum fields during its ripening stages, *i.e.,* October and November, which coincided with bamboo flowering.

Table 8.2: Rodent Species Diversity in Bamboo Flowering Areas of Mizoram.

Common Name	Scientific Name	Locality
Mackenzie's rat	*Rattus mackenziei* Thomas	Kolasib
House rat	*Rattus rattus* Linnaeus	Kolasib and South Hlimen
Himalayan rat	*Rattus nitidus* Hodgson	South Hlimen and Kolasib
Himalayan Chestnut rat	*Niniventer fulvescens* Gray	South Hlimen
Lesser Bandicoot rat	*Bandicota bengalensis* Gray	Bukvannei
Sikkim rat	*Rattus sikkimensis*	Saiha
Albino type rat	*Rattus norvegicus*	Lunglei

Implications of Bamboo Flowering

People in northeast India and elsewhere in the world, believe that bamboo flowering is the harbinger of famine. Though such famines are common in some East Asian countries like Myanmar and Japan and also in southern Africa, in the Indo-

Myanmar frontier tract, especially in Mizoram, they caused much devastation. 'When bamboo flowers, famine, death and destruction will soon follow', goes a traditional saying in Mizoram. In general, bamboo flowering is considered a bad omen. Reference to this can be found in the epic *Mahabharata,* written more than 5000 years ago. Back in 1959, bamboo flowering set off a chain of events in the state of Mizoram that ultimately led to one of the most powerful insurgencies spanning over two decades against the Indian Union.

(i) Rodent Outbreak

The popular belief is that the gregarious flowering of bamboo produces large quantities of seeds, resulting in a population explosion of rats (having short life cycles), which, in turn, leads to famine. People of Mizoram recognized two types of famines *viz., Mautam* and *Thingtam* of which, *Mautam"* is said to be much more severe. All the bamboos flower at the same time and the rat population increases suddenly over extensive area causing famine. First record of *" Mautam"* was in 1864 when rats multiplied spontaneously and hundreds of Lushais were killed as a result of famine. A similar outbreak of rats associated with bamboo flowering was also reported in 1910-1912. The flowering of *B. tulda* and *D. longispathus* during 1924 - 1928 stretched over a period of 4 -5 years causing *" Thingtam"*, with gradual buildup of rat population inducing more of scarcity than famine.

Bamboo flowering in 1958-59 resulted into an outbreak of rodent population, which spread in an area of 1000 square miles and developed into a "Plague" in Mizoram and ate up the entire food stock and standing crop. The rodent outbreak was further spread in the neighbouring area of Cachar (Assam) and Tripura where about 1000 acres of green fields were devastated resulting into famine. The rats devourd on paddy, chilli, betelnut, cotton, chestnut, fruit, millet, bamboo shoots, tobacco stem, ginger and certain grasses too. Simultaneously all the crops were eaten up resulting into famine.

Nutritive Value of Bamboo Seeds

Small rodents are likely to have a powerful reproductive response to the high nutritive value of bamboo seeds and abundant food supply during mast seeding (synchronized production of seed at long intervals by a population of plants) of bamboo, which is true for *Rattus,* a genus with native species associated with Indian and Asian mast seeding bamboo. On an average bamboo fruits are 131.86 g in weight, 11.71 cm in length and 18.58 cm in diameter. Analysis of nutrients content of *M. baccifera* fruits from Meghalaya and Mizoram showed that out of 20 amino acids only tryptophan was present and that the carbohydrate content ranged from 26.6-27.0 per cent, lipid content 6.37-9.22 per cent and protein content 8.75-9.27 per cent comparable to paddy in their protein content. Moisture content (72.84 per cent), dry matter (27.16 per cent) and ash or mineral content (3.68 per cent) of fruits were also recorded.

Comparative Nutritive Values of Bamboo Fruits and Paddy

As paddy is common food of rats in NE Region therefore, both paddy and bamboo fruits were analysed in the laboratory for their nutritive values. Analysis of nutrient contents of *M. baccifera* fruits from Meghalaya and Mizoram showed that out of 20 amino acids only tryptophan (205 mg/g N) and methionin (24.94 mg/g N) were present in traces, which were not sufficient for an increase in the reproductive behaviour. The carbohydrate and dry matter contents were quite low in bamboo fruits as compared to paddy. Whereas, lipid contents were higher in bamboo fruits but proteins and mineral contents were at par in both the materials (Table 8.3).

Table 8.3: Nutrients of Bamboo Fruits and Paddy.

Sample	Moisture per cent	Dry matter per cent	Ash per cent	Protein per cent	Carbohydrate per cent	Lipid per cent
Melocanna baccifera	72.84	27.16	3.68	9.01	26.90	7.80
Paddy	10.26	89.74	3.31	9.84	74.52	1.44

The fruits of *M. baccifera* were comparable to paddy in their protein content. Experiments on the effect of feeding of bamboo fruits of *M. baciferra* on the reproductive behaviour of rodents under laboratory conditions indicated that paddy is preferred over bamboo fruits. It was observed that when fruits of *Melocana bacifera* and paddy (as check) were fed to *Rattus nitidus* it was recorded that bamboo fruits had no obvious effect on breeding of experimental rats.

Food Preference

The consumption of bamboo fruits of *Melocanna baccifera* by rodents was believed to increase the fertility of rodents and thereby producing lots of offspring. Feeding trials conducted under laboratory conditions with bamboo fruits, other feeds and other feed with additives (mustard oil and pure ghee) showed that rodents preferred (per 100 g body weight/day) maize (9.42 g), rice (9.42 g), gram (8.64 g) with mustard oil; and maize (8.25 g), rice (8.14 g) and gram (7.59 g) without additives to bamboo fruits (1.73 g/100 g body weight/day), which reflected that the bamboo fruits were least preferred food materials (Table 8.4). No preference was observed for feeding on bamboo fruits compared to the conventional feed. Feeding behaviour and nutritive values showed that the bamboo fruit do not have any effect on the reproductive behaviour.

Feeding of maize with mustard oil resulted in the highest per cent body weight gain (16.14 per cent) in a period of 60 days, which was followed by maize (14.79 per cent), rice with mustard oil (13.32 per cent), rice (12.27 per cent), gram with mustard oil (4.23 per cent), gram (3.31 per cent) and paddy (10.21 per cent). On the other hand, feeding of bamboo fruit resulted in 15.94 per cent reduction of body weight and death in four days. Reduction in body weight of experimental rats was also recorded in case of maize with ghee additive (6.21 per cent), rice with ghee (5.08 per cent) and gram with ghee (4.23 per cent).

Table 8.4: Comparative Feeding Preference by *Rattus* sp.

Feed	Initial Rat wt (g)	Consumption per 100 g Body wt/day	Per cent Gain/Reduction in wt (after 60 days)
Maize	164.10	8.25	+ 14.79
Maize + mustard oil	174.40	9.72	+ 16.14
Maize + ghee	180.40	6.30	− 6.21
Rice	159.33	8.14	+ 12.27
Rice + mustard oil	186.83	9.42	+ 13.32
Rice + ghee	186.90	6.14	− 5.08
Gram	137.33	7.59	+ 3.31
Gram + mustard oil	157.98	8.64	+ 4.23
Gram + ghee	188.67	5.07	− 4.97
Bamboo fruit	143.67	1.73	− 15.94*
Paddy	194.65	6.50	+ 10.21

* Rats died after 4 days.

(ii) Effect on Environment

This strange phenomenon wrecks ecological havoc because of two reasons. First, the bamboo plants die after flowering and it will take at least several years for bamboo plants to re-seed, leaving bare the exposed soil which could be disastrous and also lead, to food scarcity. The second factor is that rats feed on flowers and seeds of dying bamboo trees, which is said to activate a rapid birth rate among rodents that leads to the huge rat population feeding on agricultural crops in fields and granaries and leading to starvation and death. The phenomenon of fluctuations in mammalian population, especially in rodents, has been recognized all over the world.

Most bamboos are distinct from ordinary grasses in their perennial tree-like growth habits, and flowering (and seeding) only once, at the end of very long vegetative growth phases, followed often by the death of the flowered clumps. Death of bamboo forests after gregarious flowering results in much loss and precipitates an ecological crisis. Loss of this "Green gold" will affect the life of all Northeasterners very badly.

Water, which is already a scarce resource in most of the hills, will become scarcer. Experts says that during bamboo flowering in Mizoram in 1958-59 there was a sharp rise in temperature followed by a spell of dry arid weather, which had direct fallout on the health of the people. Not only that water scarcity will also be affecting badly the life of people.

During *Mautam*, all bamboos flower and die. The immediate impact is danger of fire incidence due to huge stockpile of bamboo everywhere. The northeastern states, which have massive bamboo forests, are in particular danger. The dead plants render the land they fall on useless. Only rodents and pests thrive in the waste. Flowering kicks off a chain of dreaded events that brings famine in the area, leaves hectares upon hectares infertile and unlashes an uncontrollable population of pests on food grains and food scarcity for wild life. Due to scarcity of proper food, water and huge stocks of rotting bamboo may lead to spread of various diseases.

(iii) Effect on Rural Livelihood

A thriving economy revolves around bamboo. Presently bamboos constitute an important industrial raw material and are vital to the economy of many countries. The pulp and paper industry, construction, cottage industry and handloom, food, fuel, fodder and medicines annually consume around 22 million tonnes of bamboo. The expected mass bamboo flowering will eliminate the entire standing stock of culms within 2-3 years. Processing industries and local weavers lose their resource of raw material. If this resource left un-harvested and go waste by dying will cost a loss of around Rs. 12,000 million time/unit.

The common Northeasterners depend on bamboo for almost everything – from a raw material to build their homes to food and as one of the few resources of cash. Bamboo rotting over hundreds of hectares and growth of rat population will have a devastating effect on Jhum cultivation, on which a majority of the rural folk still depend for growing food, thus affecting the already precarious food paucity of the rural people. This major resource of income- such as Jhum field produce, the vegetables from the wild and bamboo shoots that are sold in town markets would disappear, at least for a crucial period of time, seriously affecting the sparse family budget. Bamboo is a utility plant, a gift of nature to the people from time immemorial, specially in rural areas. The development of artisan skills for handicrafts and wider utility of bamboo may provide more employment opportunities and better income distribution for the rural people.

Recently, we have observed that the local people of West Kameng district of Arunachal Pradesh and Kolasib District of Mizoram burnt down the flowering clums of bamboos because of a belief that 'flowering of bamboo heralds disaster'. This fear, coupled with superstition, has resulted in hundreds of tribal families leaving their hamlets. The bamboo flowering in different hilly areas of the region, which is not a regular phenomenon, has caused panic among the tribals who fear a possible food crisis and epidemic as already more than 500 tribals in the remote areas have been affected by gastroenteritis and viral fever.

Conclusion

It is, therefore, concluded that there is no definite relation between bamboo flowering and rodent upsurge because bamboo fruits do not contain higher nutritive value than paddy; moreover, these fruits do not contain estrogenic substances in required quantity so as to increase breeding in experimental rats.

References

Anonymous (1999). State of forest report. Published by Forest survey of India. Ministry of Environment and Forests, Govt. of India. pp. 113.

Bahadur, K.N. and Jain, S.S. (1981). Rare bamboos in India. *Indian J. Forestry.* **4**: 280-286.

Singha, I.B., Bhatt, B.P. and Khan, M.I. (2003). Flowering of *Bambusa cacharensis* Mazumder in the South part of northeast India- a case study. *J. bamboo and Ratton* **2**(1): 57-63.

Chapter 9

Regeneration of Bamboo in North-East India after Recent Large Scale Death Due to Gregarious Flowering

Ratan Lal Banik

National Mission on Bamboo Application (NMBA), Delhi, India
E-mail: bamboorlbanik@hotmail.com

Introduction

Northeastern part of India, being a part of the subtropics, has mostly clump forming bamboos both in forests and in villages. The region accounts for 28 per cent of total bamboo growing area of India and produces about 66 percent of bamboo of the country(Kulkarni and Rao, 2002). In this region alone nearly 2.31 million hectares is under bamboo and the potential availability is approximately 2.00 million tons. Within recent decades, intensive biotic interference due to increasing population pressure and urbanization, including road, dam and building construction has razed many hills and destroyed the vegetation. Furthermore, gregarious flowering among some naturally growing major bamboos *viz., Melocanna baccifera, Dendrocalamus hamiltonii, Schizostachyum dullooa* and *Gigantochloa andamanica* in the forests of the north east region has led to wide spread mortality of the bamboo. This resulted in acute shortage of raw materials, canopy openings and denudation of hills. In addition to that, a number of important bamboo species like *B.tulda, B.cacharensis, B.jaintiana* also have been dying due to sporadic flowering.

A number of common economic bamboo species like, *Bambusa balcooa, B. nutans, B.polymprpha, B.tulda, B.vulgaris* and *Thyrsostachys oliveri,* have been cultivated, mainly in the homesteads of the region and are now rapidly deteriorating in qualities and quantities due to overexploitation to meet the increasing demand. Besides, the farmers in most of the villages especially in Tripura started cultivating rubber trees in their homesteads with plantation subsidy provided by Agencies like Rubber Board thereby encroaching and replacing the bamboo stocks.

Flowering Nature

Millions of bamboo plants (Clumps) have been growing either together forming pure bamboo forest or in patches forming mixed vegetation covering vast tract of land in the region. When they are matured to flower all do not flower at a time in the same year. At the beginning, some in a population *i.e.,* about 10 per cent clumps flower scatteredly and die, known as *Initial Sporadic.* Then within 3-5 years more than 80 per cent clumps flower at a time *Gregariously* and die completely in a population. After gregarious flowering and mass scale death, the remaining scatteredly growing living clumps in the population show *Final Sporadic* flowering with simultaneous death. Such *final sporadic* flowering continues for 2-4 years till all the remaining members die.

A few records of recent flowering in some important bamboo species of the region are presented in Table 9.1.

Natural Regeneration of Bamboos in the Flowered Areas

The natural regeneration (NR) of bamboos occurs profusely after each gregarious flowering. It presents no difficulties apart from the necessity of affording adequate protection. Gregarious flowering starts at some point in the bamboo forest and gradually spreads in waves, covering the whole area up to 34 years. Bamboo seedlings are usually seen on the ground just below the flowering mother clumps. Masses of fertile seeds are shed in the immediate vicinity of the clumps and germinate profusely on the onset of rains. They are particularly numerous on bare or freshly exposed soil. Seedlings could also be seen in slopes far below the clumps where the washed away seeds accumulated during rains. Banik (1988) reported that the density of naturally occurring seedlings of *B. tulda* and *Dendrocalamus longispathus* after one to two months was about $45/100$ cm^2 in depressions and valley areas and lower on the slopes $(3/100$ $cm^2)$. There is intense competition among the seedlings themselves and as a result of *natural thinning,* gradually in three or four years time clusters begin to form and eventually in six years or more the area carries a homogeneous crop of more or less equally spaced young clumps. After 12 months it was observed that seedling density was better in thinned out than in the unthinned areas. Competition between seedlings was high in the unthinned areas and as a result the rate of mortality was higher than that in the thinned areas.

The natural regeneration process of bamboo is to be aided through different silvicultural managements for protecting wild seedlings and restocking the flowered area. The following aiding operations have been found essential for hastening the success of natural regeneration Aided Natural Regeneration or ANR and to convert the forest into a bamboo forest again (Banik 1988, Fu Mayoi and Banik 1996).

Table 9.1: Recent Flowering Records of some Important Bamboo Species of North-East India.

Species [Flowering Nature]	Locality	Recent Year of Flowering	Reference
Bambusa cacharensis [Sporadic flowering]	**Assam** (lower Assam) **Tripura**	2004-09 2003-2010	**Author** Author
Bambusa jaintiana [Sporadic flowering]	**Meghalaya**(Garohill) **Tripura (north)**	2005-06 2010	Naithani (2007) Author
Bambusa tulda [Sporadic flowering]	**Assam-Kokrajhar,** Charaikhola Guwahati **Tripura** (south, west)	2007-2008 2003, 2008-2011	Author Author
Dendrocalamus hamiltoni [flowered sporadically, then gregariously and finally sporadic]	**Arunachal** **Assam** Deopahar (Numaligarh) North Cachar Hills **Manipur**–Tamenglong **Meghalaya** (N C hills) **Mizoram**–Kolasib, Mamit **Nagaland**	2005 (Greg.) 2000-2001 2001 2000-2004 1997-98 2006-2008 2003-2005	Thakur (2005) Tripathi (2002) Yadava (2002) Banik(2004) Tripathi (2002) Author Author
D. longispathus [Initial Sporadic flowering is going on]	**Tripura** (West, South)	2010	Author
Gigantochloa andamanica [Flowered sporadically, then gregariously and finally sporadic]	**Tripura** (West, South)	2003-04 2006-08 2010	Author
Melocanna baccifera [Initially Sporadic, then Gregarious and finally sporadic flowering]	**Assam**- Halflong, Jorhat **Manipur** -Tamenglong **Meghalaya**- Garohill **Mizoram** **Tripura** (all over State)	1988, 2008 2003-05, 2007-08 2004-2008 2002-04, 2006-10 1995, 2004-09 2010	Gupta (1988), Tripathi (2002) Author Naithani (2007) Author Banik (2004, 2010) Sharma (2008)
Schizostachyum dullooa [Sporadic to Gregarious flowering]	**Assam** (Cachar) **Tripura** Hrishmukh, Gandachera) (Kailashahar, Jampuihills)	2010 2003, 2004 2010-11	Nath and Das (2010) Banik (2004) Author

Silvicultural Treatments Aiding the Natural Regeneration (ANR) Process

At the initial stage (from year 1 to 3) the wild seedlings of muli (*Melocanna baccifera*) as well as other bamboo species (from year 1 to 5) need proper care and nursing for survival in the process of obtaining successful natural regeneration (Banik1988). The following aiding operations have been found essential for hastening the success of natural regeneration (Aided Natural Regeneration) (Banik 1988, Banik 2010, Fu Mayoi and Banik 1996). The local Joint Forest Management Committee (JFMC) may be involved in all the following steps of work to obtain success at regeneration process of muli and other bamboo species.

i) Survey, Identify and Demarcate the Area, Weeding/Jungle Cleaning

First of all a survey should be done in the month of March - April to get acquainted with the topography and extent of flowered area under seeding and the boundary be demarcated. A map of the site may be drawn for systematic monitoring and evaluating the regenerating process. At the initial stage during *first year* of natural regeneration mature seeds fall randomly on the ground from the newly seeded mother clumps, germinate profusely on the onset of rains and start growing.

The seeded mother clumps are mostly dead and devoid of leaves thus allows entry of sunlight to the ground and as a result the population of weeds and vines increases on the forest floor. Many weeds and climbers start suppressing the regenerating seedlings of bamboo and as a result many seedlings die because of weed suppression and competition among many seedlings. The proper weeding including vine cutting and maintenance of the bamboo seedlings enhance the growth and within 4 – 6 years seedlings develop into merchantable clumps, whereas it takes 10 – 15 years in the natural condition if left unattended. Kurz (1876) also observed weed suppression problem for the regeneration of bamboo seedlings in Myanmar and mentioned that in unattended sites the regenerating population became thin and bamboo area got squeezed.

The condition of wild bamboo seedlings is found to be *better under partial shade* and *lower weed conditions* than under full weeds and complete shade condition (Banik 1988). In complete shade, almost all seedlings gradually degenerate. Dead standing mother clumps influence better growth of bamboo seedlings by providing partial shade. To *provide partial shade* to the regenerating bamboo seedlings felling of the dead mother clump must be delayed at least for 69 months (Banik 1988). Harvesting or burning of dead mother bamboos within 23 months of seed germination hinders the regeneration process by destroying almost all the bamboo seedlings. Therefore, *felling operation of dead bamboos in the early stage of regeneration should be discouraged* to obtain higher survival and establishment of regenerating bamboo seedlings. It has been observed that the regenerating seedlings failed to thrive well under shade, but springed up readily in the crown gaps (partial shade).

So weeding and vine cutting have to be carried out from the beginning of rainy season (May- June) in the seedling regenerating area to minimize the weed suppression. *Second weeding and vine cutting* have to be started from August and finished by September, weeds also have to be cut and taken out from the area. Simultaneously, it may be followed in the *second year*. However, at the *third year*, during June to August a thorough weeding and vine cutting have been found to be very useful for bamboo seedling growth in competition free environment.

ii) Fencing the Vulnerable Entry Side in the Area so that Cattle cannot Enter and Graze

The tender leaves and rhizomes of the bamboo seedlings are very delicious fodder for the *animals* like- cow, goats, porcupines, deer, buffalo, wild bore, etc. and thus remain vulnerable to the *grazing and trampling*. Further when animal eats up the shoots of young bamboo it pulls up the whole plant and kills it. If severe enough, *grazing is capable of wiping out* the natural regeneration of bamboos entirely. So in some vulnerable places, in the initial year during April–May, fence may be made to prevent the entry of animals for protecting the bamboo seedlings. So there is *a need of block fencing and the extent of fencing would depend on the amount of area to be protected from grazing*. Depending on the grazing intensity at the site fences may be repaired and maintained at least up to third year. However, through out the year, especially in the growing season (April–September) local JFMC members and foresters should be in vigilance to control the grazing of cattle and other wild animals.

Bamboos cultivated (example, *B.cacharensis*) in the homesteads and in the farm near the villages need special attention for protecting the seeds against the fowls, birds and the regenerating seedlings from grazing of domestic cattles.

iii) Observe and Fill Up the Gap Area by Substitute Seed Sowing/Seedling Planting at Even Spacing

During seed fall seeds are not evenly distributed throughout the regenerating area and as a result some patches of land in between remain very thinly populated or barren. *Gap planting* is usually done by substitute direct sowing of *Melocanna baccifera* seeds (2 seeds per pit) in pits during the month of June to August. Depending on the situation this operation may again be carried out in second and third year of regeneration process. For other bamboo species seedlings are to be raised in the nursery and more than 1-year old seedlings planted in the gap areas.

Therefore substitute sowing of seeds and/or seedling planting would improve the regenerating stock for obtaining adequate evenly distributed plant density during the process of regeneration.

iv) Proper Inspection and Monitoring

An *inspection path* (1-meter wide) needs to be laid out in the Aided Natural Regeneration (ANR) areas to monitor and evaluate the on-going process of regeneration. Such path is to be made by scraping the ground with spade and cutting jungle on both sides. The path is laid out, as far as possible, diagonally to the tract so that one can visit and inspect maximum portion of the area. Regular inspection is very important. At the end of rainy season, i.e. November, it is useful to make *Inspection path* for regular monitoring the regeneration performance of wild seedlings of bamboo.

v) Protection from Fire

During winter and dry months of December to April there is always a chance of fire in the forest floor that may kill the regenerating bamboo seedlings. So it is safer to lay out *fire line to control fire* and fire watcher may also be engaged. However, Inspection Path may also serve *as fire line* and protects the seedlings. The *burning* and / or *harvesting of dead culms* is detrimental to regeneration process by killing almost all bamboo seedlings. However, at the same time the presence of dead culms increases the risk of fire hazards during dry months (December-April). So firing should be prevented and harvesting of dead culms should be delayed in the early stage of regeneration (*at least up to year 3*) *to* obtain higher survival and establishment of wild bamboo seedlings. No burning or clear felling of the dead mother bamboo clumps should be allowed within one to three months of age of wild seedlings. Within this time period the seedlings develop dependable rhizome systems which help them to revive even after death of above ground shoots due to fire.

Therefore local forester has to make awareness campaign among the local hill people against the possible fire hazards, animal predation and *grazing*.

Gradually in two or three years time clusters begin to form, and eventually in fifth years or more the area carries a healthy homogeneous crop of more or less equally spaced young clumps of bamboo. Thus a new forest of muli and other bamboo species starts developing again.

Raising Artificial Regeneration (AR) of Bamboo

Seed Characters

Bamboo produces one seeded fruits with thin pericarp adnate to the seed coat, known as **caryopsis** and covered with a number of persistent glumes (Gambles 1896). The bamboo species having grain like seeds (caryopses) can tolerate desiccation (*orthodox* behaviour) while the *bacca* type *Melocanna* seeds are sensitive to desiccation (exhibit *recalcitrant*). The fleshy seeds of *M. baccifera* when stored in an airconditioned room retained viability up to 45 days, while it was only 35 days at normal room condition and prolonged further up to 60 days when stored with dry sand in jute bags (Banik 1994). The seeds of *M. baccifera* can be carried with sand in the jute bags during long distance transportation to minimise the damage and for retaining their viability. As the *M. baccifera* seeds are fleshy can be stored placing in a bag having slightly moist sand (need aeration, not desiccation), and storing the bag in air-conditioned room for 65-70 days.

Seed Germination and Longevity

Bamboo seeds possess embryos at their swollen stalk ends and therefore care should be taken to bury this portion in the soil during sowing for protecting the germinating radicles from being desiccated. Seeds should be sown in polythene bags just after collection. Seed should be sown as soon as they are collected. The swollen stalk end of seeds should be burried in the soil. The germination media (soil and cowdung 3:1) should be wet, but not waterlogged. Seeds of all the above mentioned bamboo species start germinating within 37 days of sowing and continue up to 15-25 days (Banik 1987a). Care should be taken to protect germinating radicles from being desiccated by regular watering the seedbed. Bamboo seeds germinate at a higher percentage under shade than in direct sunlight. Thus bamboo seeds can be considered as *negatively photoblastic* (Banik 1994).

In *M. baccifera* seedling survival is maximum (70-75 per cent) when raised from the seeds heavier than 50 g, but it drops (50per cent) when raised from light weight (7-16 g) seeds. Different types of abnormalities such as, rootless plumules, stunted radicles and radicles growing upward etc. are not uncommon in the seedlings produced from light weight seeds.

Seed Bed and Germination Process

Seeds be sown in Beds or Polybags as soon as collected. For germinating a few amount of seeds polybags or pot is preferred. However, in case of muli bamboo (*Melocanna bacifera*) the freshly collected seeds may be directly sown in the planting pits.

A seed bed is a small, raised and prepared patch of land where seeds are sown for germination and seedlings production. The common size for a seed bed is 5 metres long, 1.2 metres wide and 15 centimetres deep. The ideal media of germination is a mixture of sandy loam soil, sand, cow dung/FYM at the ratio of 2:2:1. The seeds should be sown in a line, rather then scattering or broadcasting. After sowing, a thin layer of the mixture should be applied across the bed.

Before sowing, seed may be soaked in water overnight. Seeds possess embryos at their swollen-end, so bury this part, *i.e.*, sow horizontally/lay down in side. The soil mixture in the seed bed should be moist and well drained.

Seedling Nursing and Management

Two to four leaved seedlings are to be pricked up from the seed bed and transplanted into polythene bags, size 15 x 23 cm, filled with soil and Farmyard Manure (FYM) at the ratio of 3:1. Adequate watering and cleaning of weeds from the beds and poly bags are to be practiced regularly.

Maintenance of Partial Shade

Initially, seedlings do best in partial shade compared to direct sunlight (Banik 1997). Complete shading over the seedling should be discouraged. The emergence of shoots is successive. The new shoots are bigger and taller than older ones. The germinating plumules are very thin (12 mm diameter) in *B.cacharensis, B. tulda, D. hamiltonii, D. longispathus, D.andamanica, S.dullooa* and thick (46 mm diameter) in *M. baccifera*. Within 14 weeks plumules elongate rapidly into stems bearing single leaves arising alternately. The stems of *B. tulda, D.hamiltonii, D. longispathus, G. andamanica, S.dullooa* and are somewhat woody in nature, but *M. baccifera* has a soft and succulent stem with vigorous growth. *M. baccifera* seedlings quickly get elongated (175 cm) and become thick (0. 8 cm, diameter) at three months of age (Banik 1994). A rhizome system starts to develop in the seedling after 12 months of germination, and at a young stage the rhizome movement is strongly geotropic.

Fertilizer Application

Application of nitrogen and phosphorus alone or in combination increases total biomass of bamboo seedling. The increase in total biomass was found to be much greater (88 per cent) due to N treatments than phosphorus treatments (24 per cent). The highest total biomass (7.31 g/plant) of *B. tulda* seedling was recorded in $N_{100}P_{50}$ as compared to 2.94g/plant in treatment $N_0 P_0$. A two-split application after 4 and 8 weeks after germination was superior to single application after 6 weeks.

Controlling Root and Rhizome Growth in Nursery

In the nurseries roots and rhizomes of seedlings penetrate the neighboring polythene bags of other individuals. This creates a mass of twisted and intermingled roots and rhizomes of seedlings. As a result, the roots and rhizomes are damaged at the time of transportation. Frequent shifting of seedlings from one bed to another helps in minimizing the root- rhizome intermingling at nursery stage. Seedlings need regular weeding and daily watering at nursery stage.

Utilization of Wild Seedlings

Wild seedlings of bamboo, look like rice or wheat seedlings and are often seen as a thick mat on the ground just below the flowering mother clumps. These densely populated seedlings compete strongly in the wilderness for survival and should be thinned out to minimise competition, and thus can be utilized in raising plantation. Wild seedlings of different bamboos can be collected and transplanted in the polythene bags and used as planting stocks.

Two to four leaved stage of wild seedlings of *Bambusa cacharensis, B. tulda, B.bambos, D.hamiltonii, D.strictus, D. longispathus, Gigantochloa andamanica,* and *Schizostachyum dullooa* are the best for collection, while in *M. baccifera* germinating seedlings (only with plumule and radicle) are best.

The rainy days (May –August) are best time for collecting wild seedlings from the forest floor. Then the seedlings have to be brought under shade net (50 per cent shade/light) for 1 week for hardening and needs watering through intermittent fogging. After 1 week increase the over head light (70 percent light) by replacing overhead shed-net for healthy growth and maintain fogging. Maintain in this way for 3-4 weeks with regular weeding. Gradually shift them under open sky.

If the seedlings are more than 15cm tall (in case of *dalu* bamboo) chop off the top, this reduces transpiring surface and thus drying of seedlings get delayed resulting less mortality.

Nine to 12 month old seedlings survive at higher rate (about 80-90 per cent) in the field after transplantation. The rainy season is the best time for planting of seedlings.

Seedling Multiplication

Since the development of *macroproliferation* method (Banik 1987b), bamboo seedlings have been multiplied through rhizome separation. Generally 5 – 9 month old seedlings of *B. bambos, B. cacharensis, B. tulda, D. hamiltonii, D. longispathus,*

D. strictus, Gigantochloa andamanica, Schizostachyum dullooa and *Thyrsostachys siamensis* are best for multiplication. Usually seedling can be multiplied to 3-5 times in number through this technique.

Steps for Macro Proliferation

1. Remove the soil to expose the rhizome of the plants. Separate the individual shoots at the rhizome neck region with the help of a secateur such that each shoot retains a portion of the rhizome system and roots.

2. Trim the upper portion of the shoots leaving two nodes, to restrict apical dominance and to enable production of more shoots.

3. Seal the cut surface with wax. Dip the cuttings in Bavistin (0.1 per cent) solution before planting them in multiplication bed or in polythene bags.

4. Replant, water and harden the each proliferated unit under shade for 3–5 days before bringing them to the nursery bed under the sun.

The growing season is the best time for macro proliferation operations.

Raising Bamboo Plantation (Artificial Regeneration or AR)

Bamboo requires gentle terrain sites, with deep, loose, and fertile sandy loam soil for good growth. Lower slopes of the hills are good planting sites; upper steep slopes have to be avoided. Bamboos do not survive under deep shade. Under sowing of seeds or seedling planting may be done in the wellthinned or widely spaced forest plantation.

On the valleys and lower gentle slopes, in the hills, planting pits of 60 cubic cm are dug at 5m x 5m or 6m x 6m spacing for cultivating big size caespitose clump forming bamboo species – *Bambusa balcooa, B. bambos, B. cacharensis, B.tulda, B. nutans, B. vulgaris, Dendrocalamus aper, D. longispathus, D. hamiltonii, D. strictus, Giantochloa andamanica, Schizostachyum dullooa,* etc. Therefore in a hectare, either 400 or 280 propagules (seedlings or cuttings) are necessary.

In the barren hills and scrubby forests close planting of *Melocanna baccifera* seedlings is suggested, and the planting pits are to be dug at 3m x 3m spacing, so 1111 pits per ha. The planting pit size may be 30 cm x 30cm x 30cm.

By Mixing Different Cohorts of Muli Bamboo

While raising a *muli* bamboo plantation efforts have be made to use the seeds from different available sources having diverse flowering duration (short flowering cycles 30-35 years, long flowering cycles 45-50 years, and others) *i.e.,* of interseeding periods (cohorts). The seeds of same cohort (having same duration of interseeding period) should not be planted continuously covering large tract of land, rather plantation be raised in patches. However, seeds or seedlings of both short and long interseeding populations may not be available always in the same year. The short duration population (30-35 years of interseeding period) started flowering earlier in late eighties. The seeds or seedlings of that short interseeding cohort already might have been planted in the forests. In the mean time the long interseeding populations

(45years) started flowering in some state and the available seeds or seedlings of this cohort may be collected and planted nearer to those short interseeding cohort. The seeds/seedlings of dissimilar cohort (flowering population) may be planted side by side in patches or blocks covering the gap areas. Thus the said locality would have *mosaic plantation* (Banik 2004) with a mixture of two or more different cohorts (flowering populations) of *M. baccifera*. In the next flowering time all these populations raised in a *mosaic plantation* will flower after different interval of times, and so all the *muli* clumps will not flower and die at a time exposing a vast tract of land and making sudden shortage of resources (Banik 2004, 2010).

The existence of diverse duration of interseeding populations would also offer possibilities of *frequent availability of seeds* in *muli* bamboo in the next flowering time. Keen observations on seeding along with localities and their documentation are important in this regard. Further, raised *M. baccifera* plantations having diverse populations would maintain a *wide range of genetic base*.

By Mixing Different Bamboo Species with Muli

Mosaic plantations of *muli* may also be raised by mixing with other local bamboo species. During 2002-2009 some bamboo species like *Dendrocalamus hamiltonii, Gigantochloa andamanica, Schizostachyum dulloa, Bambusa tulda, B. cacharensis,* etc flowered and seeded at different areas of northeast India (Assam, Nagaland, Manipur, Mizoram, Meghalaya, Tripura). All these species may be planted in small blocks mixing with *muli* bamboo on the hill tops, slopes, valleys, and riverbanks. But, specifically, the seedlings of *Schizostachyum dulloa* should be strategically planted in gullies and along the shady stream bank. Other locally available different big size clump forming useful bamboo species like *Bambusa vulgaris, B. balcooa, B. polymorpha, Thyrsostachys oliveri* are also good choice for raising *Mosaic Plantations* of *muli* nearer the human habitation (in scatteredly formed small villages/hamlets on the hills). As these species, presently, are not in seeding stage and, especially, *B.balcooa* and *B.vulgaris* do not produce seeds after flowering, can be planted with cuttings or any other vegetative planting materials. In the next flowering time this mixed mosaic bamboo vegetation will not flower and die at a time because the clumps of *B. balcooa, B. vulgaris* rarely flower and interseeding periods of others species are also not similar to *M. baccifera.*

Aftercare and Maintenance of Plantation

Grazing is also a limiting factor for survival of bamboo seedlings. Combined with fire or alone, if severe enough, it is capable of wiping it out entirely. So there is a need to make fire lines at the boundary of plantation site.

As bamboo plants grow, the clumps also expand and simultaneously canopy starts providing shade in between. So it is suggested not to plant any umbrageous wide crown evergreen timber/fruit trees as inter crop in bamboo plantation. The tree species having narrow, deciduous to semi-deciduous and light crown are usually preferred. Weeding may be avoided by cultivating the legume crops (*Cajanus cajan,* lentil, *Sesbenia* sp etc.) and trees (*Acacia, Albizzia* sp, etc.), in between the planting rows as intercrops. In order to reclaim the degraded jhum land of North-west part of

Zemabawk in Mizoram, two major economic and edible bamboo species *viz. Melocanna baccifera* and *Dendrocalamus longispathus* were planted in bamboo based agroforestry system (Jha and Marak 2004). The inter cropping of soyabean (*Glycine max*) was found to improve the soil condition and influenced the bamboo growth. The moisture content and amount of organic carbon in the degraded Jhum land increased from 12.1 to 19.1 percent and 0.315 to 0.375 percent after one year of intercropping. The yield of soyabean was also better 2284.375 kg per ha in *M.baccifera* plot and 2231.375 kg per ha in *D. longispathus* field. Additionally intercropping reduces the weeding problem and thereby cut down the plantation cost.

Fertilizer NPK may be applied in the form of Urea, Super phosphate, and Murate of Potash at the rate of 60, 40, and 40g per plant in three split doses at intervals of three months. Fertilization improves the growth of the plant by producing more elongated shoots with maximum number of leaves. Study showed that in fertilized plot and control (unfertilized) plot of *Melocanna baccifera* the mean culm height and number of leaves per plant were 1.26m, 70.7 and 1.12m, 74.2 respectively. Similarly the plant of *D. longispathus* in fertilized plots produced 1.43m, 72.0 and in control plot 1.18m, 61.3 (Jha and Marak 2004).

After each ring weeding *mulching* should be done around the base of the seedling. At the end of rainy season proper soil work has to be done around each of the planted seedling/cuttings. Immediately after that mulching is needed up to completion of annual drought period (December – March). Mulching conserves moisture in the pit and regulates underground temperature, and also checks the weed growth around the bamboo plant.

Production/Yield

As bamboo is a fast growing and quick harvesting crop, the output of the plantations will be apparent 35 years after afforestation and reforestation activities. In the case of *Melocanna* species, about 10 culms are produced per clump in the first year of harvest. Within first 5 year of plantation period on an average one clump can produce a total of 15 culms. After 7-8 year all land in 1 ha of *muli* plantation will be covered by bamboo due to the rapid expansion of all the raised 1111 clumps; and as a result each of the individual clumps could not be demarcated. On the basis of 3 years rotation felling about seven to eight thousand culms can be harvested from 1 ha of plantation. After 13-15 years of age clump productivity becomes more or less steady. The culm density in a natural forest of Mizoram, before flowering during 2005, was estimated about 30000 culms per hectare. The productivity (mostly in weight) can be improved 10-15 per cent by proper clump management for sustainable harvest.

References

Banik, R.L. (1987a).Seed germination of some bamboo species. *Indian Forester*113(8): 578-586.

Banik, R.L. (1987b).Techniques of bamboo propagation with special reference to prerooted and prerhizomed branch cuttings and tissue culture. Pp 160 169. **In:** Recent research on Bamboos (Eds. Rao, A.N., Dhanaranjan, G., Sastry, C.B.) Proc. of the international Bamboo Workshop. Hangzhou, China, IDRC.

Banik, R.L.(1988). Management of wild bamboo seedlings for natural regeneration and reforestation. Pp 92-95. **In:** Bamboos – current research (Eds. I.V. Ramanuja Rao, R. Gnanaharan, C. B. Sastry). Proc. of the Intl. Bamboo Workshop, Cochin, KFRI, India. IDRC.

Banik, R.L.(1994). Studies on seed germination, seedling growth and nursery management of *Melocanna baccifera*(Roxb) Kurz., Pp. 113-119. **In:** Bamboo in Asia and the Pacific. Proceedings of the 4th International Bamboo Workshop, Chiangmai, Thailand, 27-30 November 1991. International Development Research Centre, Ottawa, Canada; Forestry Research Support Programme for Asia and the Pacific, Bangkok, Thailand.

Banik, R. L.(1997). Growth response of bamboo seedlings under different light conditions at nursery stage. *Bangladesh Journal of Forest Science 26(2):13-18,*

Banik, R.L.(2004). Fatal flowers. Pp 6-7, World bamboo Congress, and CIBART/ INBAR Communiqué (News Letter), vol1 (1), February. New Delhi,

Banik, R.L. (2010). Biology and Silviculture of Muli (*Melocanna baccifera)* bamboo. NMBA/TIFAC, DST, GoI, New Delhi. 237p

Fu Maoyi and Banik R.L.(1996). Bamboo production system and their management. Pp18–33.**In:** Bamboo, people and the environment. Vol I. Propagation and management. *Proc.of the Intl. Bamboo Workshop and the Intl. Bamboo Congress*; Bali. Indonesia. 19-22 June,1995.

Gamble, J.S. (1896).The Bambuseae of British India. Annals of the Royal Botanic Garden, Calcutta. Vol.7, Printed at the Bengal Secretariate Press, Calcutta, London. p 133.

Gupta, K. K. (1988). Flowering and fruiting of bamboo *Melocanna baccifera* in North Cachar hills, Assam. (A letter to the Editor). *Indian Forester* 114(9): 602.

Jha, L. K. and Marak, Ch. (2004). Research Notes—Study on the growth performance of bamboo species *Melocanna baccifera* and *Dendrocalamus longispathus* along with crop (*Glycine max*) in degraded jhum land of Mizoram. *Indian Forester* 130(9): 1071- 1077.

Kulkarni, H. D. and Rao, J.S. (2002). Gregarious bamboo flowering in northeastern zone of India–Strategies for sustainable utilization of bamboo resources. Pp 42-46. **In:** Proceedings of expart consultation on strategies for sustainable utilization of bamboo resources subsequent to gregarious flowering in the northeast. 24-25 April. 2002. RFRI, Jorhat, Book No-12-2002., India.,UNIDO.

Kurz, S. (1876). Bamboo and its use. *Indian Forester.* 1(3): 219269 and 1(4): 355362.

Naithani, H.B. (2007). *Survey report on the distribution of bamboo species in Meghalaya,* India. Department of Forest and Environment, Government of Meghalaya. Shillong. June 2007.

Nath. A. J. and Das, A. K. (2010). Gregarious flowering of a long-lived tropical semelparous bamboo *Schizostachyum dullooa* in Assam (Correspondence). *Current Science,* Vol. 99 (2): 154-155.

Sharma, D. K. (2008). Status of bamboo resource development and its utilization in Tripura, TFD, (Tripura JICA Project-TFIPAP), NTFP CE, Hatipara, Agartala. p218.

Thakur, Azad, N. S. (2005). Study report of bamboo flowering in East Kamong dist.– Official correspondance. 17 th May 2005, Division of Entomology., ICAR Research Complex for NEH Region (Meghalaya).

Tripathi, K. C. (2002). Gregarious bamboo flowering in northeastern region and some management considerations. Pp131-135. **In:** Proceedings of expart consultation on strategies for sustainable utilization of bamboo resources subsequent to gregarious flowering in the northeast. 24-25 April. 2002. Rain Forest Research Institute (RFRI), Jorhat, Book No-12-2002., India.,UNIDO.

Yadava, M. R. (2002). Flowering of bamboo and management issues arising out of it, with special reference to *Melocanna baccifera* (Muli bamboo) in the southern parts of Assam. Pp 28-33. **In:** *Proceedings of expart consultation on strategies for sustainable utilization of bamboo resources subsequent to gregarious flowering in the northeast.* 24-25 April. 2002. RFRI, Jorhat, Book No-12-2002., India.,UNIDO.

Chapter 10

Development of Bamboo Mosaic Virus as a Plant Expression Vector for Functional Study of Genes Involved in Bamboo Flowering

Na-Sheng Lin

Institute of Plant and Microbial Biology, Academia Sinica

Introduction

Historically, several generations of farmers had noticed bamboo mosaic disease in bamboo plantations, and the frequently occurring disease has also been suggested as one of the causes of bamboo flowering. Until 1974, 2 species of bamboo (*Bambusa multiples* and *B. vulgaris*) were reported to be infected with Bamboo Mosaic Virus (BaMV), the causal agent of bamboo mosaic disease, in Brazil. Later, 2 other bamboo species in Taiwan (*B. oldhammi* and *Dendrocalamus latiflorus*) were reported to be infected with BaMV (Lin *et al.*, 1979). The virus is now known to be transmissible by mechanical injury of plants and seems to be unrelated to bamboo flowering. However, bamboo is normally propagated vegetatively, and thus, the use of nonindexed, infected plants as propagation materials has greatly aided in the spread of the disease. The infection has been found in many countries — China, Philippines, Japan, India, Thailand, Australia, France and America (Lin *et al.*, 1993; Hsu and Lin, 2004) — although these areas have incurred no severe economic loss. The number of infected bamboo species has been expanded to 14 in 7 different genera, including the most economic bamboo species for the production of bamboo shoots, *B. oldhamii* and

D. latiflorus (Lin *et al.,* 1993). BaMV usually causes severe mosaic symptoms in leaves and brown streaking in the shoots or culms, which severely lowers the market values of bamboo shoots. The infection rate in many bamboo plantations in Taiwan has been over 80 per cent. According to a Tainan Agricultural Experimental Station report (Yeh *et al.,* 1993), infection with BaMV can cause approximately 50 per cent yield loss of bamboo shoot production. No natural vector has yet been reported.

Molecular Biology of BaMV

BaMV was first isolated from infected *B. oldhamii* (BaMV-O). After the genome of BaMV was completely sequenced, in 1994, there has been tremendous progress in our understanding of the genome properties and the replication events of this virus (Lin *et al.,* 1992; 1994). BaMV is a member of the potexvirus group and has a single-stranded, positive-sense RNA genome of approximately 6.4 kb with a 94-nt 5′ untranslated region (UTR) and a 142-nt 3′ UTR with a poly(A) tail. It contains five open reading frames (ORFs). ORF1 encodes a 155-kDa protein with methyltransferase, helicase and polymerase domains and is involved in viral RNA replication (Tsai *et al.,* 2005). ORFs 2, 3 and 4, the overlapping triple gene block proteins (TGBp1-3), are involved in cell-to-cell movement (Lin *et al.,* 2004). ORF5 encodes a 25-kDa capsid protein (CP) with multiple functions: encapsidation, cell-to-cell movement and symptom determination (Lan *et al.,* 2010). Two major subgenomic RNAs, of 2.0 and 1.0 kb, are 3′ co-terminal for the expression of TGBp1 and CP proteins in infected cells (Lin *et al.,* 1992). Structural and functional analysis of the 3′ UTR has revealed that a sufficient length of a poly (A) tail is required for the formation of a pseudoknot for efficient replication of BaMV RNA (Tsai *et al.,* 1999). The conserved motif ACC/UUAA in potexviruses and carlaviruses is also found in the 3′ UTR of BaMV (Tsai *et al.,* 1999).

Several complete sequences of BaMV isolates and partial sequences have been deposited in the NCBI databank (Lin *et al.,* 1994; Yang *et al.,* 1997). About 10-12 per cent variation in nucleotide sequence is found among isolates, with ORF1 being the most conserved and the CP gene the most divergent (Yang *et al.,* 1997).

Full-length cDNA Cloning

The full-length cDNA clone of BaMV-O, pBL, was first constructed in a pUC119 vector under control of the T7 promoter (Tsai *et al.,* 1999). The *in vitro* transcripts derived from pBL resulted in infection in the protoplasts and plants of *N. benthamiana* and *Chenopodium quinoa*. With this key progress, we can manipulate the BaMV genome at the molecular level to understand the functions of viral genes and noncoding sequences. Because BaMV-O causes asymptomatic infection and accumulates a low titer of virions in infected plants, BaMV-S was selected among the collected variants followed by serial passages in *N. benthamiana*. Subsequently, the infectious full-length cDNA of BaMV-S was constructed and available under the control of the 35S or T7 promoter (Lin *et al.,* 2004). The virus produced 5- to 10-fold higher virus titers than did BaMV-O in infected plants and thus, was suitable for development as a virus-based plant expression vector.

Development of BaMV as a Plant Expression Vector

Over the last 2 decades, plant viruses have been developed as vectors for expressing foreign gene proteins, polypeptides, and nucleotide sequences or for inducing gene silencing in infected plants (Scholthof *et al.,* 1996; Faye and Gomord, 2010). Successful viral vectors include DNA and RNA viruses such as Pararetroviruses, Geminiviruses, Bromoviruses, Tombusviruses, Tobamovirus, and Potexvirus (Scholthof *et al.,* 1996; Faye and Gomord, 2010). Compared to expressing foreign proteins in transgenic plants, this system offers a number of advantages: (1) a higher copy number expression of foreign genes, (2) simpler and quicker expression, (3) no requirement for selection and regeneration of stable transformation lines, (4) easy manipulation in the different hosts as long as they are hosts of the virus, and (5) less impact on the ecological environment. Similarly, a stronger plant expression vector could trigger a stronger gene silencing event in infected plants.

Several strategies are used to express foreign genes by viral vectors in plants. For BaMV, we successfully expressed foreign proteins and peptides by gene insertion (Lin *et al.,* 2004) and epitope presentation (Yang *et al.,* 2007). By duplicating the subgenomic RNA promoter upstream of the CP gene, the green florescent protein (GFP) gene can be engineered into the BaMV vector (BaMV-GFP) and substantial GFP protein expressed in infected plants (Lin *et al.,* 2004). By epitope presentation, the VP1 peptide of foot-and-mouth disease virus was fused to the CP gene of BaMV. The VP1 peptide could be exposed on the surface of chimeric BaMV virions by immunogold labeling and thus induce protective immunity in swine (Yang *et al.,* 2007). This BaMV-based vector technology may be applied to other vaccines for correct antigen expression.

BaMV Viral Vector for Studies of Gene Functions Involved in Bamboo Flowering

Bamboo has a relatively long vegetative stage. The switch from the vegetative to reproductive stage must be controlled by multiple pathways and signals. Although the bamboo genome sequence is not available, more than 10,000 full-length cDNA sequences from *Phyllostachys heterocycla* cv. *pubescens* have been reported (Peng *et al.,* 2010). Moreover, approximately 4,500 expressed sequence tags from flower buds of *B. oldhamii* have been sequenced, of which about 1,500 are flower-specific unigenes (Lin *et al.,* 2010). These flowering-related genes include Zinc finger protein, MADS protein, MYB transcription factors, and F-box protein (Lin *et al.,* 2009; 2010). However, these genes have not been functionally characterized. Because no stable transformation system is available for bamboo plants, BaMV could provide an ideal system for overexpression and knockdown assays of genes involved in bamboo flowering. The candidate genes can be inserted into the BaMV genome under the duplicated subgenomic RNA promoter for overexpression or silencing. As well, because of the unpredictable flowering time of bamboo, the transient assay can be conducted in other monocotyledonous plants, such as barley or rice, that BaMV can infect. Information on bamboo flowering is limited, so these BaMV-based vectors should have considerable promise for large-scale functional analysis of cDNA clones of bamboo flower-related genes.

References

Faye, L., and Gomord, V. (2010). Success stories in molecular farming- a brief overview. *Plant Biotechnol J* 8, 525-528.

Hsu, Y. H., and Lin, N. S. (2004). Bamboo mosaic associated to the complex: BaMV-satBaMV. 2004. *In* "Viruses and virus disease of Poaceae (Gramineae) (Ed. by H. Lapierre, P.-A. Signoret)". INRA, Paris. 725-726.

Lan, P., Yeh, W. B., Tsai, C. W., and Lin, N. S. (2010). A unique glycine-rich motif at the N-terminal region of *Bamboo mosaic virus* coat protein is required for symptom expression. *Mol Plant-Microbe Interact* 23, 903–914.

Lin, E. P., Peng, H. Z., Jin, Q. Y., Deng, M. J., Li, T., Xiao, X. C., Hua, X. Q., Wang, K. H., Bian, H. W., Han, N., and Zhu, M. Y. (2009). Identification and characterization of two bamboo (*Phyllostachys praecox*) AP1/SQUA-like MADS-box genes during floral transition. *Planta* 231, 109-20.

Lin, M. K., Chang, B. Y., Liao, J. T., Lin, N. S., and Hsu, Y. H. (2004). Arg-16 and Arg-21 in the N-terminal region of the triple-gene-block protein 1 of *Bamboo mosaic virus* are essential for virus movement. *J Gen Virol* 85, 251-259.

Lin, N. S., Chen, M. J., Kiang, T., and Lin, W. C. (1979). Preliminary studies on bamboo mosaic disease in Taiwan. *Taiwan Forestry Research Institute Bulletin* No. 317, 1-10.

Lin, N. S., Lin, B. Y., Lo, N. W., Hu, C. C., Chow, T. Y., and Hsu, Y. H. (1994). Nucleotide sequence of the genomic RNA of bamboo mosaic potexvirus. *J Gen Virol* 75, 2513-2518.

Lin, N. S., Chai, Y. J., Huang, T. Y., Chang, T. Y., and Hsu, Y. H. (1993). Incidence of bamboo mosaic potexvirus in Taiwan. *Plant Dis.* 77, 448-450.

Lin, N. S., Lin, F. Z., Huang, T. Y., and Hsu, Y. H. (1992). Genome properties of *Bamboo mosaic virus*. *Phytopathology* 82, 731-734.

Lin, X. C., Chow, T. Y., Chen, H. H., Liu, C. C., Chou, S. J., Huang, B. L., Kuo, C. I., Wen, C. K., Huang, L. C., and Fang, W. (2010). Understanding bamboo flowering based on large-scale analysis of expressed sequence tags. *Genet Mol Res.* 9, 1085-93.

Peng, Z., Lu, T., Li, L., Liu, X., Gao, Z., Hu, T., Yang, X., Feng, Q., Guan, J., Weng, Q., Fan, D., Zhu, C., Lu, Y., Han, B., and Jiang, Z. (2010) Genome-wide characterization of the biggest grass, bamboo, based on 10,608 putative full-length cDNA sequences. *BMC Plant Biology* 10, 116.

Sainsbury, F., Liu, L., and Lomonossoff, G. P. (2009). Cowpea mosaic virus-based systems for the expression of antigens and antibodies in plants. *Methods Mol Biol* 483, 25-39.

Scholthof, H. B., Scholthof, K. B., and Jackson, A. O. (1996). Plant virus gene vectors for transient expression of foreign proteins in plants. *Annu Rev Phytopathol* 34, 299-323.

Tsai, C. H., Cheng, C. P., Peng, C. W., Lin, B. Y, Lin., N. S., and Hsu, Y. H. (1999). Sufficient length of a poly(A) tail for the formation of a potential pseudoknot is required for efficient replication of bamboo mosaic potexvirus RNA. *J Virology* 73, 2703-2709.

Tsai, C. H., Meng, M., Line, N. S., and Hsu, Y. H. (2005). The replication of bamboo mosaic virus and its associated RNA. *In* "Recent Advances in RNA Virus Replication (Ed. by Kathleen L. Hefferon)", pp. 265-282.

Yeh, C. C., Cheng, A. H., and Hwang, H. Y. (1993). Indexing of bamboo mosaic virus and propagation of virus-free bamboo. *In* "Proceedings of the Symposium on Plant Virus and Virus-like Disease (Ed. by Chiu RJ, Yeh Y.)". Council of Agriculture, Plant Protection Series, No. 1, pp. 275-282.

Yang, C. D., Liao, J. T., Lai, C. Y., Jong, M. H., Liang, C. M., Lin, Y. L., Lin, N. S., Hsu, Y. H., and Liang, S. M. (2007). Induction of protective immunity in swine by recombinant *Bamboo mosaic virus* expressing foot-and-mouth disease virus epitopes. *BMC Biotechnol* 7, 62.

Chapter 11

Effects of Taiwan Vole (*Microtus kikuchii*) Herbivory on Yushan Cane (*Yushania niitakayamensis*) Growth in Alpine Meadows of Taiwan

Y. Kirk Lin[1,2], Su-Han Yeh[1], and Jih-Tay Hsu[3]

[1] *Institute of Ecology and Evolutionary Biology*
[2] *Department of Life Science*
[3] *Department of Animal Science and Technology*
National Taiwan University, Taipei, Taiwan

Introduction

Plant-animal interaction is one of the central issues in community ecology. It has been well documented that insects affect the growths of plant populations by actions such as pollination, dispersing seeds, transmitting diseases, and herbivory (Wagner 1997, Hummel *et al.*, 2009, Rosumek *et al.*, 2009). Similarly, mammals could affect the growths of plant populations by above actions as well as altering soil nutrients and microbial communities (Stuart-Hill and Mentis 1982, Zavada and Mentis 1992, Borghi and Giannoni 1997, Ritchie *et al.*, 1998, Gomez-Garcia *et al.*, 1999, Gomez-Garcia *et al.*, 2004, Feeley and Terborgh 2005, Darabant *et al.*, 2007, Gough *et al.*, 2008). Therefore, animals could not only negatively influence the recruitment, growth, and survival rate of plants, they benefit plants as well (Huntly 1991).

Herbivory is often considered to have negative effects on plants. For example, in the Qinling Mountains of China, basal diameters of new shoots, and clonal regeneration of culms of arrow bamboos, *Fargesia qinlingensis*, were significantly less in giant panda herbivory plots compared to control plots (Wang *et al.*, 2007a, Wang *et al.*, 2007b). In response to the foraging of herbivores, plants evolved counter strategies (Stuart-Hill and Mentis 1982). Some produced deterrents, which provide mechanical or chemical protection (Alonso-Diaz *et al.*, 2008). The physical property (Hudson *et al.*, 2008), secondary metabolites (Alonso-Diaz *et al.*, 2008), as well as nutrient contents (Bergeron and Jodoin 1987, Willig and Lacher 1991, Derting and Hornung 2003, Parsons *et al.*, 2006, Morrison and Hik 2008) of plants could determine how much a plant is consumed by herbivores, which usually prefer individuals or parts with high proteins (Bergeron and Jodoin 1987, Deguchi *et al.*, 2001), low fibers (Deguchi *et al.*, 2001), and low secondary metabolites (Gomez-Garcia *et al.*, 1999, Gomez-Garcia *et al.*, 2004, Alonso-Diaz *et al.*, 2008). Some plants tolerated foraging by reallocating the biomass between roots and shoots, or increasing productivity (Ritchie *et al.*, 1998). Some responses could in fact increase the fitness of plants. For example, in the Spanish Pyrenees, the density, asexual reproduction, and seedling abundance of a geophyte, *Merendera montana*, were higher in vole-active plots than vole-excluded plots. The voles' burrowing activities would increase the spreading of seeds, seedling, and asexual buds (Borghi and Giannoni 1997, Gomez-Garcia *et al.*, 1999, Gomez-Garcia *et al.*, 2004). Thus, herbivores seemed to have positive effects on the plant.

Yushan cane (*Yushania niitakayamensis* (Hayata) Keng f.) is a perennial monocarpic species classified as *Bambusoideae*. The phenology of Yushan cane has been documented by several researchers (Table 11.1). For example, Chen (1997) conducted a field survey at the He-Huan mountain, and found Yushan cane produced new shoots from April to June, grew leaves from July to September, and some leaves withered from October to March. The sexual reproduction of Yushan cane is likely mass synchronous flowering and seeding, like most other bamboo species. Liao (2004) reported a mass flowering event in Snow Mountain during August~November 2001. There has been no periodicity of flowering recorded thus far, however. The Yushan cane uses rhizome ramets for asexual reproduction, which has been classified as metamorph II, running rhizome with sympodial culms (Lin, 1976).

Generally, the growth of bamboos can be divided into two stages: the first-year shoots, which are unbranched, covered in sheaths and the >1-year culms, which are branched, lignified, without sheaths attached at the nodes (Tripathi and Singh 1994, Widmer 1998). Temperature and humidity are the two main factors that limit the production of shoots. An increase in temperature and humidity of soil would lead to early shooting (Wang and Kao 1986). Bamboo shoot farmers in Taiwan maintained bamboo fields by keeping the humidity of soil high, plowing soil frequently, removing old rhizomes, and fertilizing. Harvesting emerging shoots in proper ways would lead to a secondary shooting in the same year (Liu *et al.*, 2009).

Chen (1983) showed that the distribution of biomass of different Yushan cane parts changed with phenology. He separated the Yushan cane into four parts: rhizome, culm (main stem with side branches removed), side branch (leaves removed), and leaf. The relative weight of rhizomes (per cent total weight per unit area) was at the highest point (~ 38 per cent) a month (early February) before shooting (*i.e.*, new shoots emergence) started in March, declined rapidly until shooting ended (~ 20 per cent) in mid May, and remained constant until late July (~ 20 per cent) when it started

Table 11.1: The Phenology of Yushan Canes Observed in this Study Compared to those Reported in Previous Studies, including Chang's Observations at WangHsiang Area, DaSyue Mt., KuanWu area (Chang 1981), Chan's observation at Chu-Tung area (Chan 1983), Chan's observation at Mt. He-Huan (Chan 1997), and Liao's (2004) observation at Snow Mt.

	Month											
	1	2	3	4	5	6	7	8	9	10	11	12
Chang(1981)	Leaves wither			New shoots emerge			Growth peak			Leaves wither		
Chan(1983)	Growth terminate		New shoots emerge			Growth peak				Growth terminate		
Chan(1997)	Leaves wither			New shoots emerge			Growth peak			Leaves wither		
Liao(2004)	Leaves wither			New shoots emerge			Leaves growth and flowering				Leaves wither	
My observation	Little growth and leaves wither			New shoots emerge			New shoots grow taller			Growth slow down and leaves wither		
				Old culms grow branches and leaves			Old culms grow branches and leaves					

to increase gradually. The relative weight of culms remained fairly constant (~45 per cent) throughout the year, except a peak in late July (~61 per cent). The peak coincided with the end of active growing season of new shoots, thus indicated that the increase in culm weights came from the growth of new shoots. The relative weight of side branches was bimodal, with peaks (~20 per cent) occurred in early February and late July, respectively. The relative weight of leaves remained constant (~10 per cent) throughout the year. The seasonal pattern of biomass distribution observed by Chen (1983) suggested that Yushan canes started transfer of energy to shooting in early spring (early February). New shoots started to emerge in March, stopped in May, yet continued to grow tall without branching until late July~August. The increase of side branches and new leaves on old culms started in May, peaked in June or later, and declined. New shoots started to grow side branches and leaves in August until pass late October.

The nitrogen concentration of different Yushan cane parts also changed with phenology (Chang 1981). The nitrogen concentrations of leaves were 2~6 times those of culms and rhizomes throughout the year. Higher nitrogen concentration occurred in August~October and March~May for leaves and rhizomes/culms, respectively. Yushan cane was the most palatability food for Taiwan voles in the He-Huan mountain (Ho 2009). With seasonal variation in nutrient contents, Yushan canes could be consumed by voles on different parts in different season, which could have different effect on Yushan canes. For example, Chan (1997) suggested that depredation on emerging shoots by rodents could reduce the number of nodes, and thus the overall height of Yushan canes.

Research Questions

The current research wants to know how the herbivory by voles affect Yushan cane growth. The consumption of culms and leaves could decrease photosynthesis efficiency, and retard Yushan cane growth. Yet, it may increase light penetration to and increase the amount of aboveground litter on the ground level, which might improve conditions in terms of temperature and humidity that facilitate Yushan cane growth. The consumption of different parts could be affected by nutrient contents (Figure 11.1).

We hypothesize that —

1. Taiwan voles have feeding preference on different parts of Yushan canes.
2. Feeding preference can be explained by nutrient contents of different parts: highly preferred parts have higher protein and lower fiber contents than less preferred parts.
3. Consumption by Taiwan voles has positive effects on the growth of Yushan canes by increasing light penetration and aboveground litter.

Materials and Methods

Study Area

The field study was conducted in an alpine meadow (24°08′36.4″N, 121°17′17.4″E, 3007~3070 m in altitude) at He-Huan Mountains of the Taroko

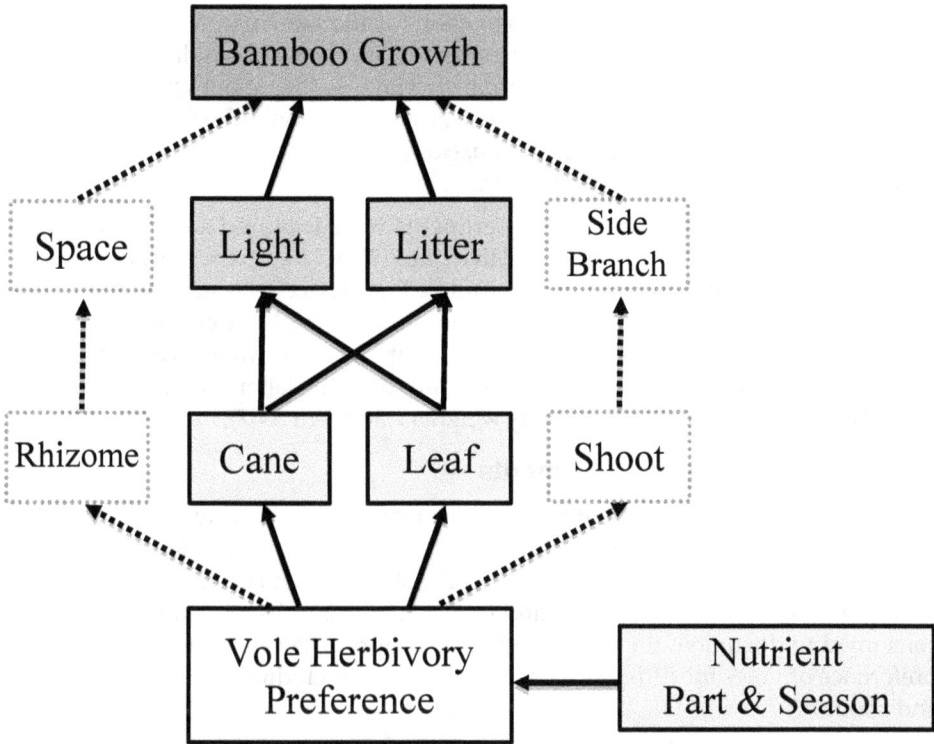

Figure 11.1: The Concept Map of the Current Study.
The consumption of different Yushan cane parts could change the availability of underground space and penetrating light from canopy, and influence the amount of ground litter accumulated and side branches produced.

National Park. The average temperature is 7.0°C and rainfall 366 mm (Ho 2009). The Yushan cane (*Yushania niitakayamensis* (Hayata) Keng f.) was the dominant plant species in the meadow. The Taiwan vole (*Microtus kikuchii*), whose primary and preferred food was Yushan cane, was the dominant small mammal in the meadow (Lin and Lin 1989, Ho 2009).

Laboratory Experiments and Analyses

We used two feeding experiments in the laboratory to determine the preference of different Yushan cane parts by the Taiwan voles. The experiments, described in detail below, were conducted in the High-Altitude Station of the Institute for the Endemic Species Research, about 5 km from the mountain meadow where field experiments were performed. We conducted nutrient analyses in a laboratory (the laboratory of Dr. Jih-Tay Hsu's) in the Department of Animal Science and Technology, National Taiwan University. All voles and plant samples used in the laboratory experiments and chemical analyses were collected from a meadow near the field site at the entrance to the Mt. Cilai, about 500 m from the mountain meadow where field experiments were performed.

Based on the phenology of Yushan canes at the He-Huan Mountains (Chen 1997), we performed the laboratory experiments in three seasons: January (during non-growing season, November~February), May (during shoot-growing season, March~June), and September (during leaf-growing season, July~October) for two years, January 2008 ~ October 2009. Because of September typhoons, the September experiments were postponed till October.

The Taiwan voles used in the experiments were captured a day before each experimental trial using Ugglan Special live traps ($25 \times 7.8 \times 6.5$ cm^3), baited with roll oats mixed with peanut butter. A ball of crumpled newspaper was supplied to provide shelter. The captured voles were housed individually in plastic cages ($50 \times 25 \times 20$ cm^3) with 10 cm thick wood shaving, and supplied with water, oats, and sweet potatoes for at least six hours to allow voles adapt to the laboratory environments. We only used adult animals, with body weights > 30g (Wu 2007), in the experiments.

Feeding Preference Experiments

Two types of feeding preference experiments were conducted: bite trials and cafeteria trials, in sequence. The bite trials tested the preference of voles for aboveground Yushan canes: shoots or culms with leaves. The shoots had no leaves in May, but had a few leaves on the tips in October and January. We also observed how canes were consumed to document the consumption behaviors. The cafeteria trials tested the preference of voles for different Yushan cane parts, including rhizome, culm, leaf, and shoot.

In 2008, we used nine voles in January, seven in May, four in June, and ten in October. In 2009, We used three voles in January, March, and May each, and two in July, and October each.

Bite Trials

The bite trials were set up to mimic the way voles would encounter Yushan canes in the wild. During each trial, for each replicate, ten >1-year live culms and ten first-year shoots were arranged in a 5 x 4 checkerboard pattern by inserting the bases of cut culms or shoots into a $10 \times 5 \times 5$ cm^3 (LxWxH) wet flower-arrange foam. The foam was then placed in a $30 \times 20 \times 15$ cm^3 (LxWxH) plastic cage fenced in with 1-cm mesh, 60-cm high mesh wires. A Taiwan vole was introduced into the cage at 8 pm, provided with 15 g of sweet potato and oat, each (control food). At the 4th (midnight) and 12th (8 am) hours, we counted the numbers of culms and shoots bitten by voles. A culm or shoot was bitten by voles if over one-third of circumference was gnawed. The bite trials were performed in 2008 only.

Cafeteria Trials

During each trial, for each replicate, 10 g of Yushan cane rhizome, culm, leaf, and shoot each, and 15 g of sweet potato and oat each were arranged in a $50 \times 25 \times 20$ cm^3 (LxWxH) cage. Each Yushan cane part was wrapped in wet paper towels in a $7.8 \times 7.8 \times 2$ cm^2 (LxWxH) plastic dish to reduce water loss. The spatial arrangement of Yushan cane parts was randomized for each replicate. A Taiwan vole was introduced into the cage at midnight. After 12 hours (at noon the next day), the vole was removed, and the left-over was collected and weighed. To estimate weight loss of plant parts due to dehydration, an experimental control was created during the same period of time by placing the same amount of food with similar set up in a separate cage without voles (Ho 2009). The weights of left-over Yushan cane parts were then adjusted for weight loss due to dehydration. The cafeteria trials were performed in both 2008 and 2009.

We calculated Manly's Alpha (Krebs 1999) to quantify the preference of each vole for each plant sample (i) in a 12-hr trial.

$$\text{Manly's Alpha } (\alpha_i) = \ln(p\text{-}_i) \, D \, \Sigma \ln(p\text{-}_i)$$

$$p\text{-}_i = T_C \, D \, T_O$$

$$T_C = T_L \times (C\text{-}_O \, D \, C_L)$$

where,

$p\text{-}_i$: Proportion of plant sample left unconsumed

T_C: Weight of plant sample left unconsumed after 12 hrs. adjusted for water loss

T_O: Weight of plant sample offered to vole

T_L: Weight of plant sample left unconsumed

C_O: Weight of control plant sample in the beginning of trials

C_L: Weight of control plant sample after 12 hrs.

Nutrient Content Analyses

In 2008 and 2009, we collected Yushan cane parts from the field when the cafeteria trials were conducted to analyze their nutrient contents. All collected samples were weighed, and kept in plastic bags to prevent water loss. They were temporarily stored in a 7 ! refrigerator before they were freeze dried, ground, and stored in -20 ! within a week. We measured five components of nutrient contents, including water, ash, neutral detergent fiber (NDF), acid detergent fiber (ADF), and crude protein (CP) of each Yushan cane part using standard methods: water (AOAC 2000), ash (AOAC 2000), NDF (Vansoest *et al.*, 1991), ADF (Goering and Vansoest 1970), and CP (AOAC 2000). Detail descriptions of methods are given in the Appendix. We also calculated the amount of hemicellulos by substracting ADF from NDF.

Field Experiments

We used two field experiments: exclosure experiment and canopy-litter manipulation experiment, to determine the effects of Taiwan vole consumption on the growth of Yushan canes. The experiments, described in detail below, were conducted in the mountain meadow nearby the Songsyue Lodge. The exclosure experiment tested the overall effects of Taiwan vole exclusion on the growth of Yushan cane, and the canopy-litter manipulation experiment simulated the effects of increased litter and decreased canopy caused by vole consumption on the growth of Yushan canes.

Exclosure Experiment

We set up a pair of 2×2 m^2 vole exclosures at six sites on the mountain meadow in December 2007. Each pair contained a vole-proof exclosure, and a leaky exclosure serving as a control treatment. Each exclosure was constructed by 1-cm mesh meshwire extended 80-90 cm aboveground and 30-40 cm deep belowground holding in place with PVC pipes staked to the ground. A 25-cm wide transparent plastic film was fixed to the top edges of the meshwire on the vole-proof exclosure to prevent voles from entering by climbing. The leaky exclosure served as a control treatment had large openings on the ground level that allowed voles enter freely. Small mammal traps placed inside vole-proof exclosures during periodic trapping indicated there was no vole. Every four month from May 2008 to October 2009, we randomly selected four 20×50 cm^2 long transects within each exclosure to census the number of culms and shoots of Yushan canes.

Canopy-Litter Manipulation

To simulate the foraging of the voles on the Yushan cane growth, we selected 12 plots with 100 per cent Yushan cane cover in the meadow randomly. In each plot, we set up a trio of different treatments. The treatments, each 50×50 cm^2 in size, were: (1) canopy reduction – we removed several culms and attached leaves to reduce ~50 per cent foliage cover; (2) litter removal – we removed the majority (>90 per cent) of leaf litter; (3) control — no culm, leaf or litter were removed. The three treatments in each plot were within 2 meters of each other. Every four month from May 2008 to October 2009, we maintained the treatments, and randomly selected two 25×25 cm^2 areas in each treatment in each plot to count the number of culms and shoots of Yushan canes. The amount of litter produced by Yushan canes every four month from litter removal plots (N=12) was oven-dried in 60! for 48 hrs and weighed.

Field Shoot Survey

We marked 99 shoots of Yushan canes in a 5×5 m^2 area on the meadow in June 2008. We recorded the height of each shoot. Every six month, we surveyed the conditions of marked shoots to see if they were consumed by herbviore.

Statistical Analyses

For the bite trials, we counted the numbers of >1-year culms and first-year shoots that were bitten by voles. Results from May and June trials were combined to represent the shoot growing season. Results from different individuals in a season were pooled together. We examined the differential consumption using a 2 x 3 (part x season) contingency table, followed by a chi-square test in each month as post hoc comparisons. The bite trials were performed in 2008 only.

For the cafeteria trials, we calculated Manly's α for each Yushan cane part for each vole tested. We examined the feeding preference of different parts using two-ways (part x season) fixed-factor ANOVAs, followed by Scheffe's tests as post hoc comparisons. Results from different individuals in a season served as replicates. The cafeteria trials were performed in 2008 and 2009, yet we only used results from 2008 in this analysis because of extremely small sample sizes in 2009.

To determine the nutrient contents that explained the feeding preference (Manly's α), we calculated the average of Manly's α of different individuals in a season to find a feeding preference value for a given month. Individual nutrient contents were first analyzed separately with simple linear regressions to show the effects of individual nutrient contents. We then used stepwise selection technique in a multiple-regression model to identify key nutrients in explaining the preference. We used results from both 2008 (January, May, June, and October) and 2009 (January, March, May, July, and October) in these analyses.

In both exclosure and canopy-litter manipulation experiments, we first calculated the ratio of >1-year culms to first-year shoots in different seasons. We examined the difference between treatments using logistic regressions (Allison 2005).

The amounts of leaf litter produced every four months in the canopy-litter manipulation experiment were analyzed using an ANOVA with blocking to test the differences among the 12 plots (blocks) and 3 seasons, followed by Scheffe's tests as post hoc comparisons (Shen 2005).

None of the data sets violated the assumptions, such as normality, of statistical tests performed. We used SAS version 9 (SAS 2003) to perform all statistical analyses. A p-value < 0.05 was accepted as statistically significant for differences.

Results

Laboratory Experiments and Analyses

Feeding Preference Experiments

Bite Trials

We counted the numbers of >1-year culms and first-year shoots of Yushan cane bitten by Taiwan voles in January, May, and October 2008. The 2 x 3 (part x season) contingency tables were significant for both the 4hrs (Chi-square test, $\chi^2=32.65$, $p < 0.0001$, Figure 11.2 A) and 12hrs ($\chi^2=21.53$, $p < 0.001$, Figure 11.2B) tests, thus indicated significant interactions. The Taiwan voles significantly preferred first-year shoots over >1-year culms during the first 4 hrs in May ($\chi^2=59.51$, $p < 0.001$) and October ($\chi^2=21.78$, $p < 0.001$), and over the whole 12 hrs periods in May ($\chi^2=67.50$, $p < 0.001$) and October ($\chi^2=37.24$, $p < 0.001$).

During the bite trials, when voles consumed the branches and leaves of >1-year culms, they would clip the leaves off the branches at the leaf petioles, and ate from the petiole ends of leaves. They often did not consume the whole leaf, and discard the leaf tips, thus produced litter on cage floors. Furthermore, the voles showed seasonal differences in their foraging behaviors on first-year shoots. In May, the shoot growing season, the voles would fell the shoots at proximately 3~5 cms from the ground, gnaw off the exterior, and eat the interior of shoots. In October and January, the leaf-growing and non-growing seasons, respectively, the voles would feed on the leaves on the culms and tips of first-year shoots. They either climbed or felled the stems to do so.

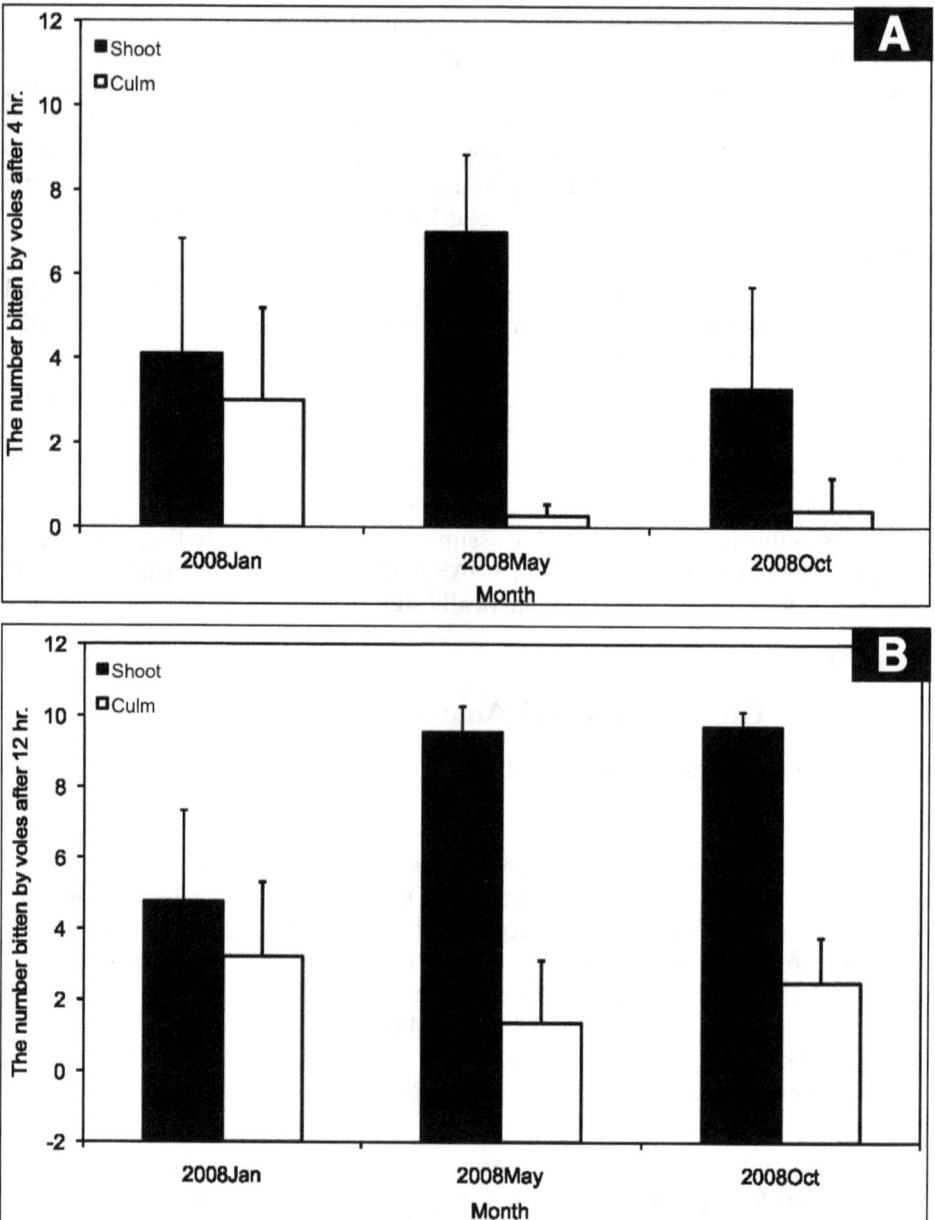

Figure 11.2: The Ratio of the Numbers of >1-year Culms and First-year Shoots Bitten by the Taiwan Voles during (A) the first 4 hrs., and (B) Over the whole 12 hrs. of Bite Trials in Three Seasons. The error bars give 95 per cent CI.

Cafeteria Trials

We used the same voles used in bite trials to examine the feeding preference of different parts (rhizome, culm, leaf, and shoot) of Yushan canes to voles in 2008.

Because there was no significant difference between sexes in feeding preference (Three-way ANOVA, part x season x sex, sex effect, $F_{1,107}$=0.62, p = 0.43), data were pooled and analyzed with a two-way ANOVA (part x season). The feeding preference of Yushan cane parts differed among seasons (Two-way ANOVA, part x season interaction, $F_{6,108}$=24.25, p < 0.001). The ranks of feeding preference were shoot > leaf > culm = rhizome in May, and leaf > shoot = culm = rhizome in January and October (Table 11.2).

Nutrient Content Analyses

The nutrient contents of different Yushan cane parts (rhizome, culm, leaf, and shoot) were analyzed in all months and years when cafeteria trials were performed. All nutrient contents measured, including ash, NDF, ADF, hemicellulose (NDF minus ADF), and CP, were expressed as percentage of dry matter weight (per cent DM), except water which was expressed as percentage of fresh weight (per cent FW), and were presented in Table 11.3 and Figures11.3–11.8. Each nutrient content measured was significantly related to the feeding preference of different parts of Yushan cane, when nutrients were analyzed separately with simple linear regressions (Table 11.4). CP (r^2=0.73, p < 0.001), ash (r^2=0.60, p < 0.001), hemicellulose (r^2=0.42, p < 0.001), and water (r^2=0.18, p < 0.01) were positively, while ADF (r^2=0.75, p < 0.001) and NDF (r^2=0.67, p < 0.001) were negatively related to feeding preference.

We followed up the simple linear regressions with a multiple regression with stepwise selection to identify key nutrient contents that explain feeding preference. Since NDF was the combination of hemicellulose and ADF, we didn't include NDF into the multiple regression model. The results indicated ADF, CP, and ash were the key nutrient contents. The final regression model gave: Preference = 0.60149 " 0.011 ADF + 0.017 CP " 0.017 ash (Table 5, r^2=0.79, p < 0.001).

Field Experiments

Exclosure Experiment

Although the exclosures were established in December 2007, and first surveyed for shoot-culm ratios in January 2008, new shoots did not emerge until March 2009, it would be inappropriate to use the ratios in January 2008 as baselines. We used the ratios in May 2008 as baselines instead. The shoot-culm ratios remained relatively constant in vole exclosures, while the ratios increased over time in control treatments (Figure 11.9). The ratios in May 2008 was significantly different from May (Logistic Regression, treatment x time interaction, $\div^2$ = 18.24, p < 0.001) and October (Logistic Regression, treatment x time interaction, $\div^2$ = 16.29, p < 0.001) 2009. Detail test results were given in Table 11.6.

Canopy-Litter Manipulation

Reducing canopy cover had a significant effect, the shoot-culm ratio of Yushan cane increased over time (Figure 11.10), using the ratio in May 2008 as baseline. The ratio was higher than that of control in May (Logistic Regression, treatment x time interaction, χ^2 = 22.67, p < 0.001) and October (χ^2 = 7.11, p < 0.01). Reducing leaf litter also had a significant effect, the shoot-culm ratio of Yushan cane increased over time

Table 11.2: The Feeding Preference of Yushan Cane Parts, Expressed as Manly's α, in different Seasons. The values give mean±1se. Different lower case letters denote significant differences based on pair-wise posthoc comparisons using Scheffe's tests. The posthoc tests were first performed for each month separately, and then combined.

	Rhizome[c]	Culm[c]	Leaf—[a]	Shoot[b]
January[a]	0.0365±0.0100[b]	0.0064±0.0047[b]	0.3948±0.0775[a]	0.0248±0.0143[b]
May[b]	0.0072±0.0065[c]	0.0080±0.0033[c]	0.2655±0.0328[b]	0.4996±0.0593[a]
October[a]	0.0446±0.0132[b]	0.0037±0.0022[b]	0.4084±0.0465[a]	0.0100±0.0046[b]

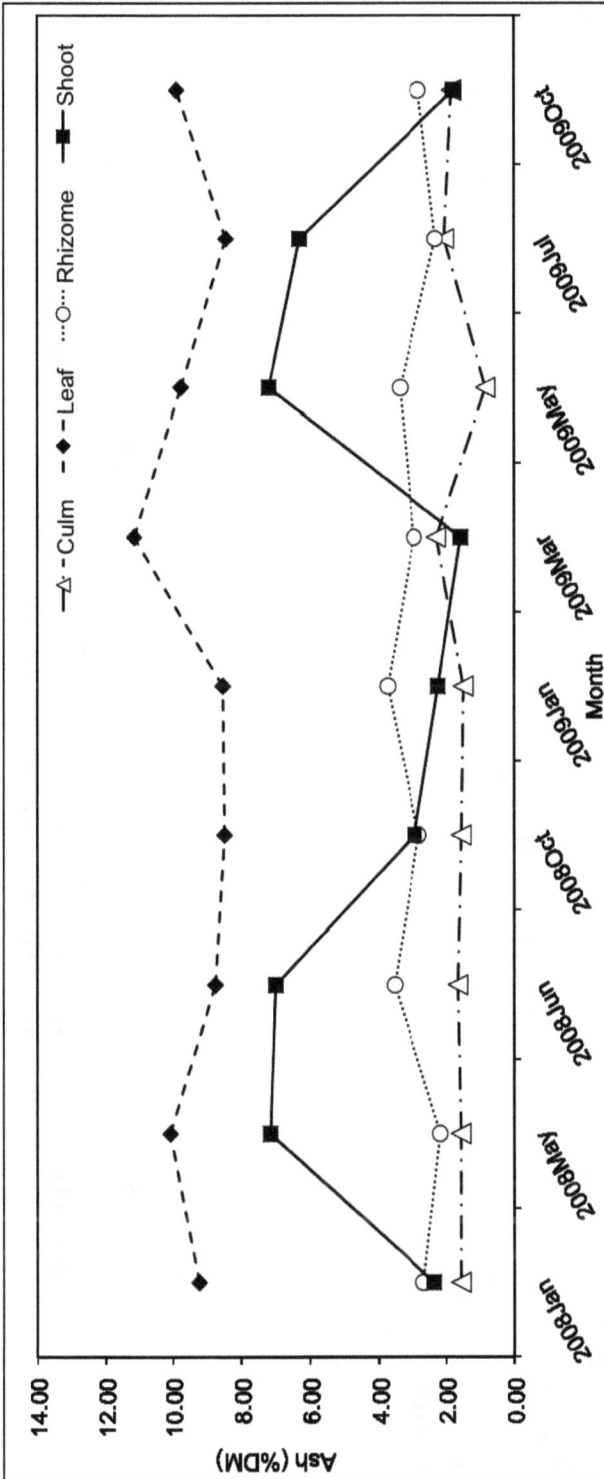

Figure 11.3: The Amount of Ash (in per cent dry matter weight) in different Yushan Cane Parts in different Seasons and Years.

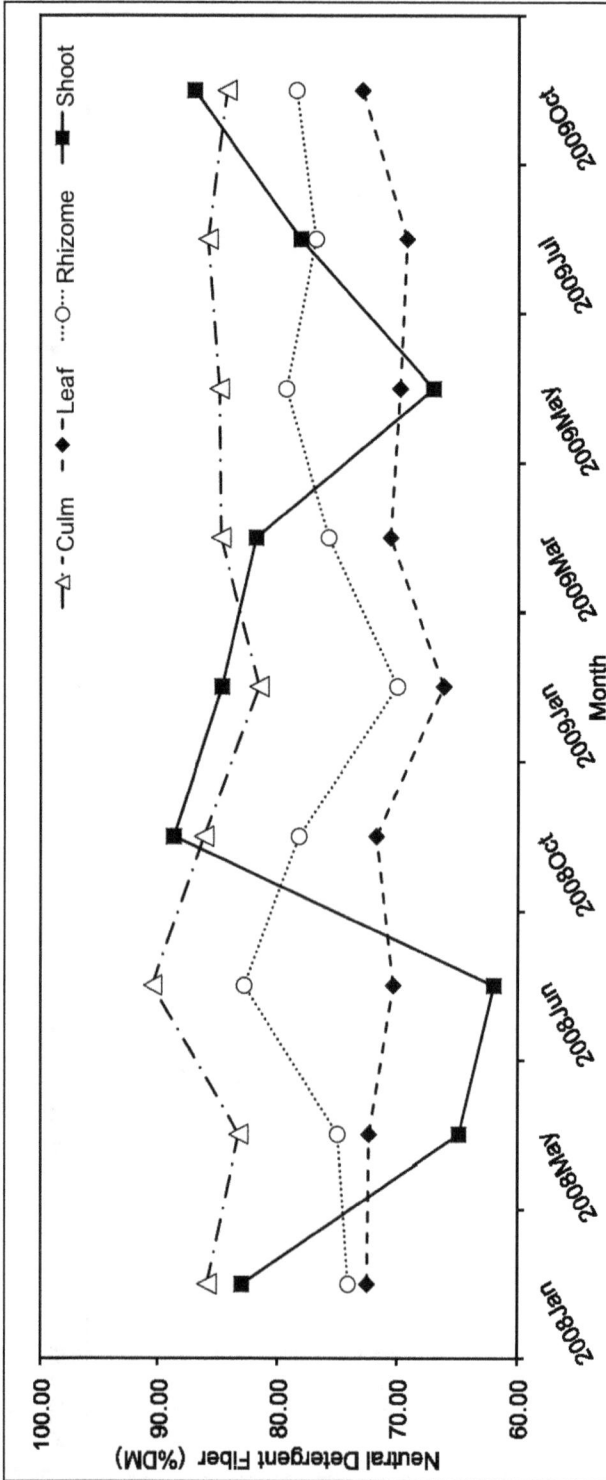

Figure 11.4: The Amount of Neutral Detergent Fiber (in per cent dry matter weight) in different Yushan Cane Parts in different Seasons and Years.

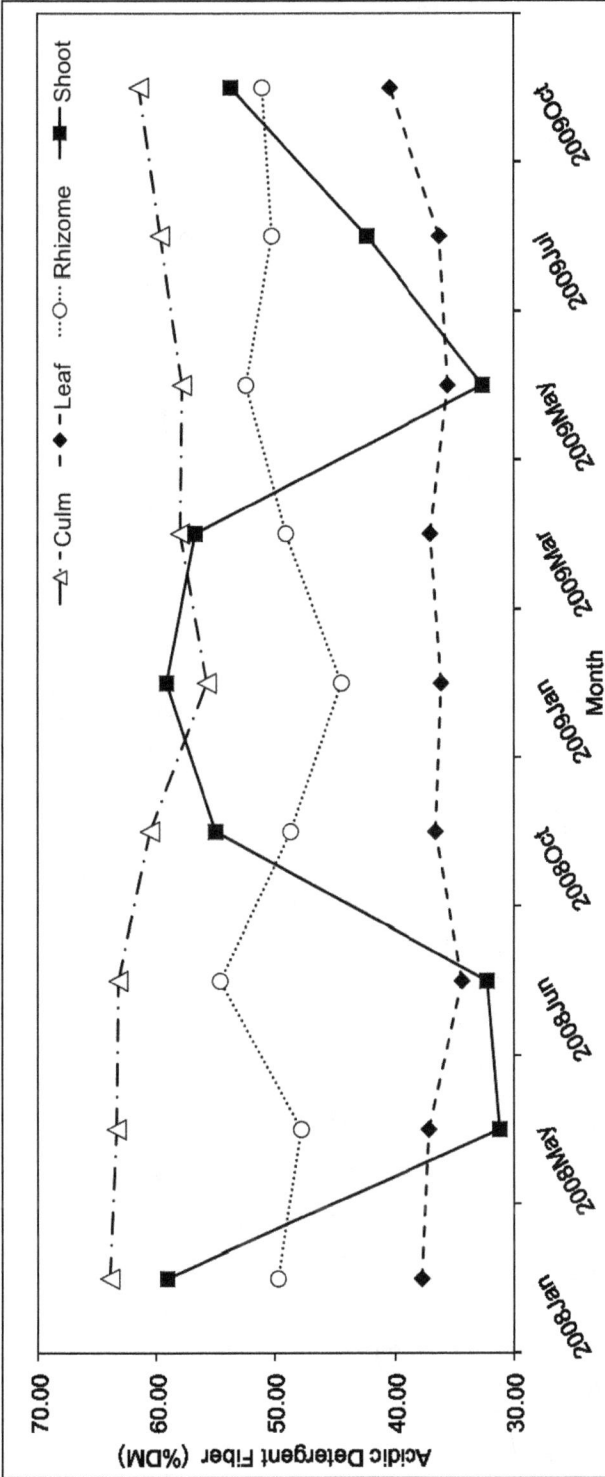

Figure 11.5: The Amount of Acidic Detergent Fiber (in per cent dry matter weight) in different Yushan Cane Parts in different Seasons and Years.

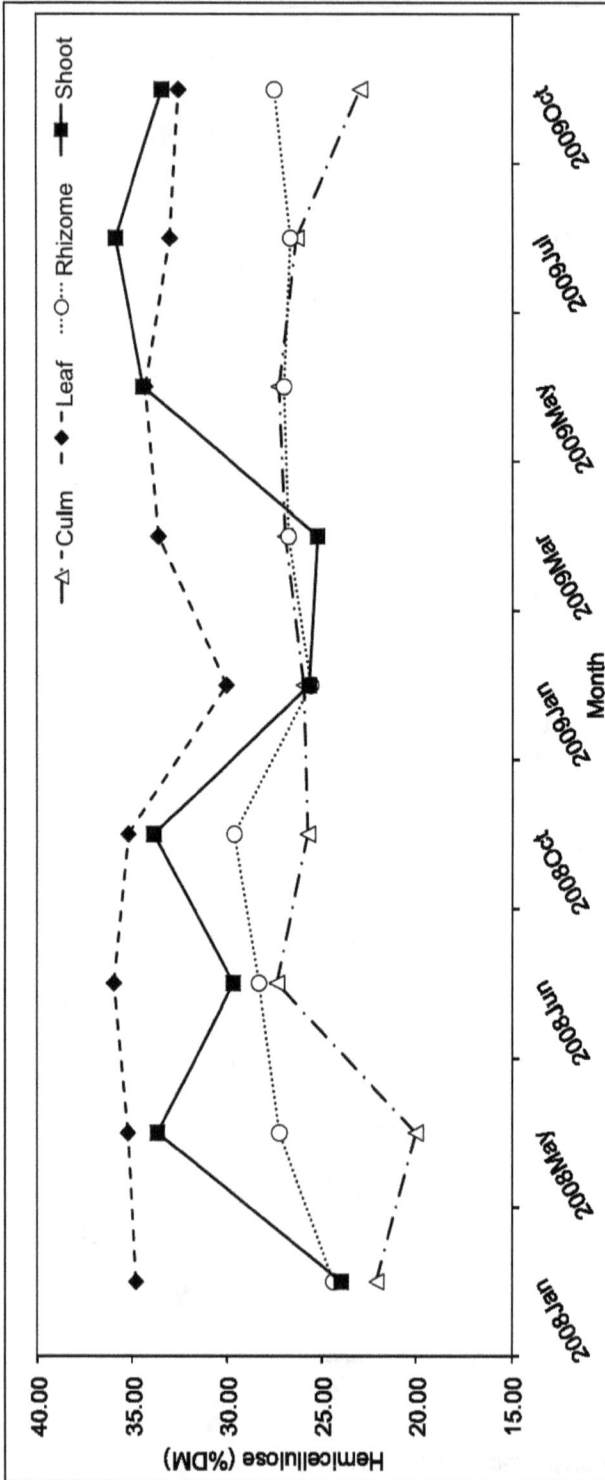

Figure 11.6: The Amount of Hemicellulose (in per cent dry matter weight) in different Yushan Cane Parts in different Seasons and Years.

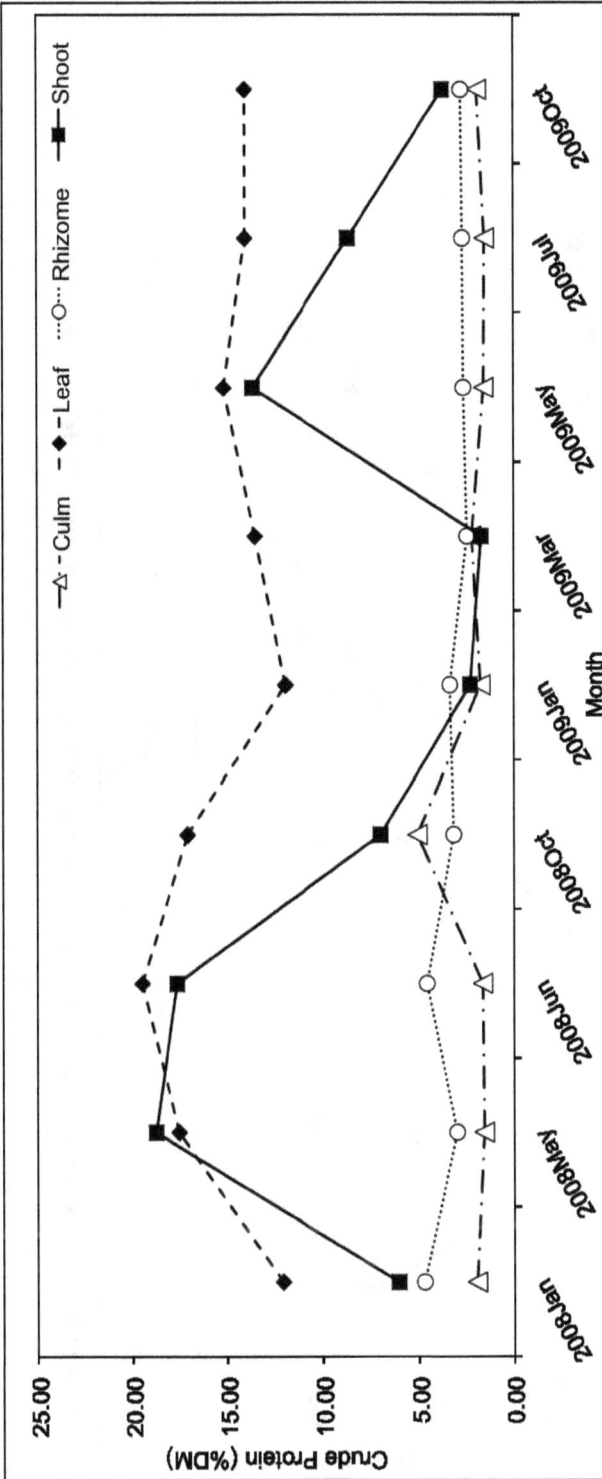

Figure 11.7: The Amount of Crude Protein (in per cent dry matter weight) in different Yushan Cane Parts in different Seasons and Years.

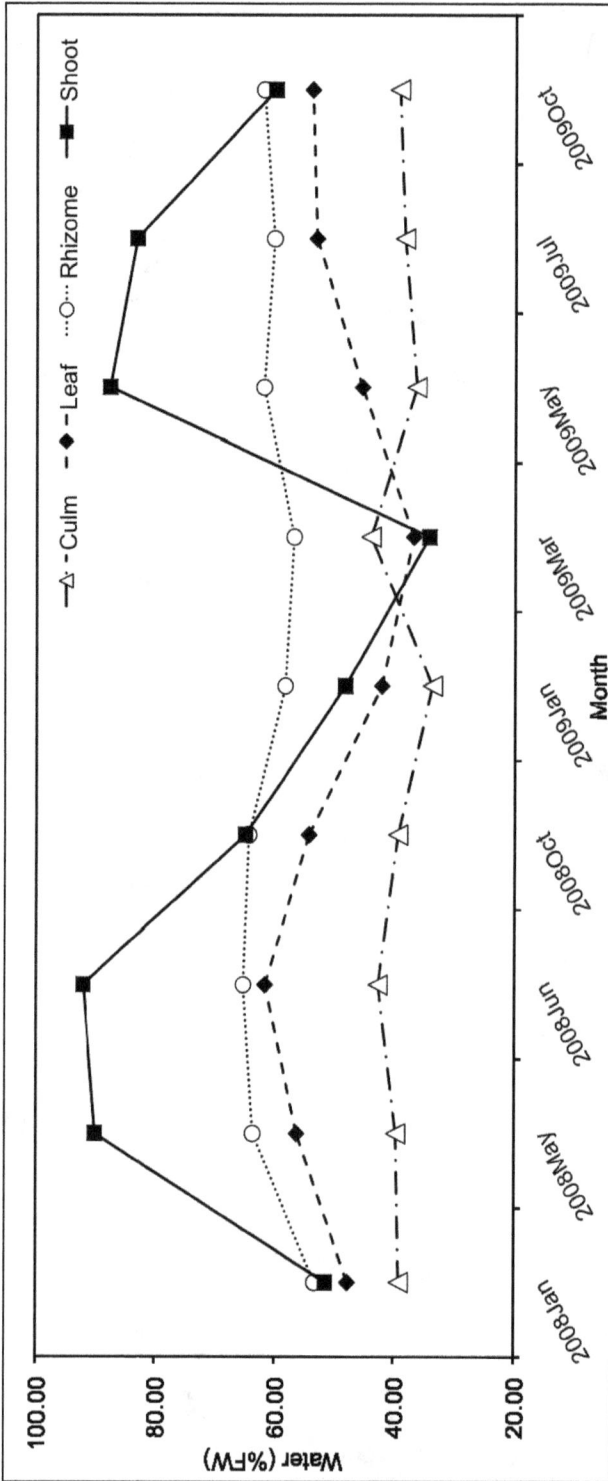

Figure 11.8: The Amount of Water (in per cent fresh weight) in different Yushan Cane Parts in different Seasons and Years.

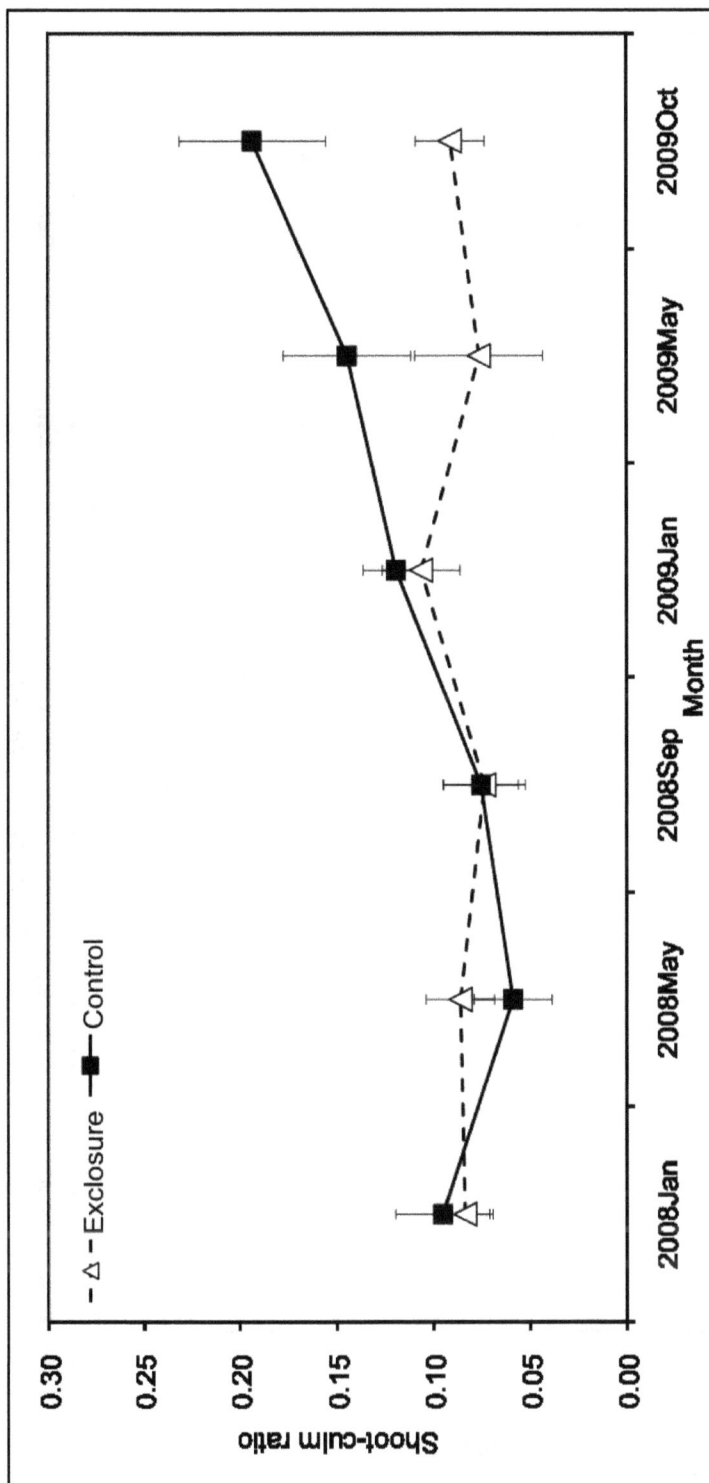

Figure 11.9: The Shoot-Culm Ratio of Vole-Exclusion and Control Treatments in different Seasons and Years. The error bars give 95 per cent CI.

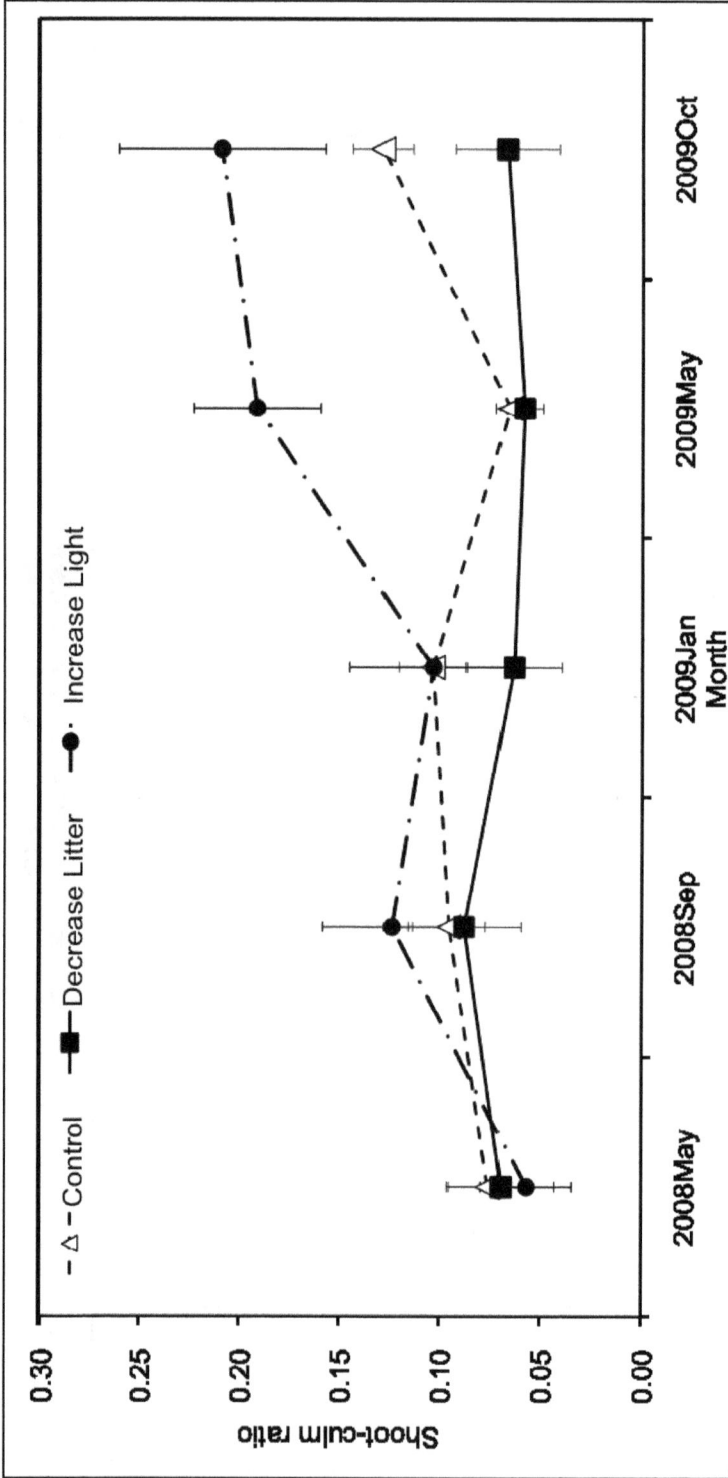

Figure 11.10: The Shoot-culm Ratio of Decrease Canopy, Decrease Litter, and Control Treatments in different Seasons and Years. The error bars give 95 per cent CI.

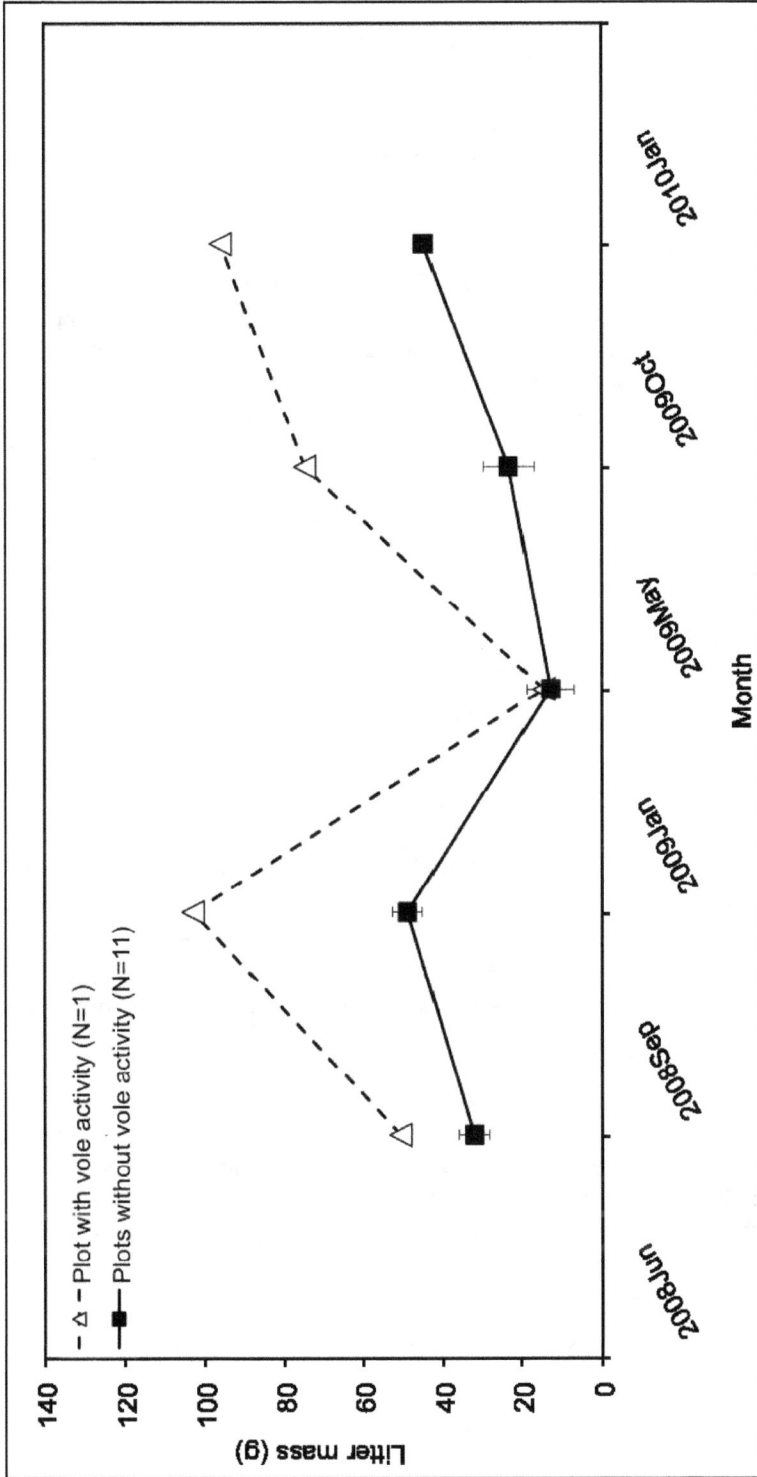

Figure 11.11: The Amount of Yushan Cane Litter Produced per 50- by 50-cm Square Area per Time Period in different Years. The Index of vole density was also given. The error bars give 95 per cent CI, N = 12.

(Figure 11.10), using the ratio in May 2008 as baseline. The ratio was higher than that of control in October (Logistic Regression, treatment x time interaction, $\chi^2 = 3.81$, $p = 0.05$). Detail test results were given in Table 11.7. The mount of litter produced differed season/year (RCBD, season/year effect, $F_{4,44}=30.11$, $p < 0.001$). One of the plots (plot D) had high vole activity, as abundant Yushan cane clipping and vole feces were present year round. Plot D did have greater amount of litter than other 11 plots (RCBD, plot effect, $F_{11,44}=5.59$, $p < 0.001$).

Field Shoot Survey

The 99 first-year shoots marked in a 5×5 m^2 plot in June 2008 had an average height of 28±11 cm (mean±1sd). Over seven months, only 4 shoots had evidence of being bitten by voles, and the height of shoots became 38±11 cm in January 2009. After a year, only 8 shoots had evidence of vole herbivory, and the height of shoots became 39±11 cm (mean±1sd).

Discussion

The voles did show significant feeding preference for different parts of Yushan canes in different seasons. The ranks of preference were shoot > leaf > rhizome = culm in May (shooting season), and leaf > shoot = rhizome = culm in January (non-growing season) and October (leaf growing season), as revealed by the bite (Table 11.1) and cafeteria (Table 11.2) experiments. The differential preference was influenced by the seasonal changes in nutrient contents of different Yushan cane parts (Table 11.3, Figures 11.3–11.8). First of all, simple linear regressions showed the amount of each nutrient content measured correlated with feeding preference significantly (Table 11.4). A multiple regression with stepwise selection further indicated the best predictive nutrient contents for the feeding preference was ADF, CP, and ash, in decreasing importance (Table 11.5). The ADF alone explained 75 per cent variation of feeding preference, while adding CP only improve the explanatory power to 78 per cent, and adding ash could improve the explanatory power to 79 per cent.

In fact, the differential preference matched very well the seasonal changes in ADF and CP. Generally, the amounts of ADF were culm > rhizome > leaf > shoot in May (shooting season), and culm = shoot > rhizome > leaf in January (non-growing season) and October (leaf growing season). The amounts of CP were leaf = shoot > rhizome = culm in May, and leaf > shoot = rhizome = culm in January and October. The leaves of Yushan canes had the lowest ADF and highest CP among all parts almost year round except shooting season. Coincidently, voles preferred to forage on leaves almost year round except shooting season. During the shooting season, when the first-year shoots with the lowest ADF and highest CP among all parts newly emerged from underground, voles showed significant preference for shoots, and leaves remained voles' second preferred food.

Several previous studies also showed that food with low ADF was considered high-quality for voles (Goldberg *et al.*, 1980, Bergeron and Jodoin 1987). Though some food had high nutritious contents, voles would not prefer it due to high fiber contents (Goldberg *et al.*, 1980). Taiwan voles seemed to prefer food with ADF contents lower than 40 per cent (dry mass, Figure 11.5). The feeding behaviors of voles further

Table 11.3: The Nutrient Contents of different Yushan Cane Parts and the Preference for different Parts by Taiwan Voles in 2008 and 2009.

Month	Part	Preference (Manly's α)	Ash (per cent DM)	NDF (per cent DM)	ADF (per cent DM)	Hemicellulose (per cent DM)	CP (per cent DM)	Water (per cent FW)
2008-Jan	Culm	0.0064	1.57	85.95	63.86	22.08	1.95	39.11
	Leaf	0.3948	9.23	72.50	37.71	34.79	12.09	47.71
	Rhizome	0.0365	2.68	74.09	49.70	24.39	4.72	53.28
	Shoot	0.0248	2.37	83.00	59.03	23.97	6.06	51.43
2008-May	Culm	0.0082	1.57	83.29	63.29	20.01	1.57	39.64
	Leaf	0.2102	10.07	72.35	37.13	35.22	17.52	56.29
	Rhizome	0.0114	2.18	75.00	47.76	27.24	2.99	63.64
	Shoot	0.5948	7.14	64.84	31.20	33.64	18.75	90.07
2008-Jun	Culm	0.0076	1.66	90.50	63.12	27.38	1.64	42.70
	Leaf	0.3622	8.75	70.32	34.35	35.97	19.47	61.69
	Rhizome	0.0000	3.53	82.83	54.52	28.31	4.57	65.27
	Shoot	0.3329	6.98	61.95	32.26	29.69	17.62	92.13
2008-Oct	Culm	0.0037	1.56	86.24	60.48	25.76	5.07	39.37
	Leaf	0.4084	8.48	71.76	36.56	35.20	17.06	54.31
	Rhizome	0.0446	2.82	78.26	48.63	29.63	3.16	64.36
	Shoot	0.0100	2.94	88.74	54.89	33.84	7.00	64.96
2009-Jan	Culm	0.0008	1.50	81.60	55.64	25.96	1.74	33.63
	Leaf	0.3658	8.53	66.13	36.09	30.05	11.95	42.14
	Rhizome	0.1981	3.73	70.01	44.41	25.59	3.35	58.37
	Shoot	0.0000	2.24	84.75	59.06	25.68	2.29	48.25

Contd...

Table 11.3–Contd...

Month	Part	Preference (Manly's α)	Ash (per cent DM)	NDF (per cent DM)	ADF (per cent DM)	Hemicellouse (per cent DM)	CP (per cent DM)	Water (per cent FW)
2009-Mar	Culm	0.0017	2.29	84.84	57.89	26.94	2.20	43.99
	Leaf	0.1483	11.14	70.59	37.00	33.59	13.52	36.81
	Rhizome	0.0601	2.96	75.78	49.02	26.77	2.45	56.88
	Shoot	0.0251	1.58	81.86	56.62	25.23	1.70	34.17
2009-May	Culm	0.0464	0.84	84.99	57.72	27.27	1.56	36.35
	Leaf	0.2815	9.75	69.80	35.52	34.28	15.14	45.48
	Rhizome	0.0206	3.36	79.34	52.33	27.01	2.63	61.96
	Shoot	0.5309	7.17	67.01	32.62	34.39	13.65	87.67
2009-Jul	Culm	0.0000	2.05	85.93	59.58	26.35	1.51	38.34
	Leaf	0.4344	8.43	69.23	36.22	33.01	14.03	53.11
	Rhizome	0.0031	2.32	76.85	50.18	26.67	2.68	60.25
	Shoot	0.1025	6.29	78.14	42.29	35.85	8.67	83.11
2009-Oct	Culm	0.0014	1.84	84.32	61.35	22.98	1.91	39.25
	Leaf	0.1942	9.88	72.94	40.39	32.56	14.03	53.83
	Rhizome	0.0106	2.83	78.49	50.98	27.51	2.76	61.93
	Shoot	0.0079	1.79	87.02	53.60	33.41	3.73	60.03

supported the key role of fibers. During the bite trials in May, the shooting season, the voles would fell the shoots at base close to the ground, gnaw off the exterior, and eat the interior of shoots. In October and January, the voles would only feed on the leaves on the culms and tips of first-year shoots. When feeding on leaves, they would clip the leaves off the branches at the leaf petioles, ate from the petiole ends, and discard the leaf tips. These behaviors suggested voles were avoiding fibers as much as they could. In this discussion, we used ADF as a synonym of fibers.

Table 11.4: The Relationship between Feeding Preference of Yushan Cane Parts and their Nutrient Contents Based on Simple Linear Regressions, N = 36 for all analyses. The levels of significance were denoted as ** p<0.01, *** p<0.001.

Nutrient Variable	β	Adjusted-R^2	$F_{1,34}$
Ash	0.0426	0.5959	52.62***
NDF	−0.0194	0.6749	73.67***
ADF	−0.0147	0.7548	108.74***
Hemicellulose	0.0266	0.4230	26.66***
CP	0.0247	0.7342	97.68***
Water	0.0052	0.1853	8.86**

Table 11.5: The Relationship between Feeding Preference of Selected Yushan Cane Parts and their Nutrient Contents Based on a Stepwise Multiple Regression, N = 36. The levels of significance were denoted as ** p<0.01, **** p<0.001.

Nutrient Variable	Parameter Estimate		Regression Model	
	β	t-value	Adjusted-R^2	$F_{3,32}$
Ash	−0.017	−1.53	0.7929	45.66****
ADF	−0.011	−3.41**		
CP	0.017	2.87**		

Table 11.6: Statistical Results of Logistic Regressions Comparing the Shoot-Culm Ratios between the Vole-Exclusion Treatment and Control. The numbers in the cells give Chi-square values. The parentheses give p-values.

Month	Plot	Treatment	Year	Treatment x Year
May 2008 vs. May 2009	15.73 (0.0077)	2.07 (0.1503)	8.82 (0.0030)	18.24 (<0.001)
May 2008 vs. Oct. 2009	2.93 (0.7110)	0.87 (0.3510)	21.20 (<0.001)	16.29 (<0.001)

The preference of Taiwan voles was less affected by CP. Previous studies had shown that food with high per cent of CP was considered high-quality (Bergeron and Jodoin 1987) and a criterion of food selection (Goldberg *et al.,* 1980, Harju and Hakkarainen 1997) for voles. The amount of ash came in third as the determinant of

Table 11.7: Statistical Results of Logistic Regressions Comparing the Shoot-Culm Ratios between the Canopy-Litter Treatment and Control. The numbers in the cells give Chi-square values. The parentheses give *p*-values.

Month	Reduce Canopy vs. Control				Decrease Litter vs. Control			
	Plot	Treatment	Year	Treatment x Year	Plot	Treatment	Year	Treatment x Year
May 2008 vs. May 2009	11.77 (0.3810)	9.66 (0.002)	13.23 (<0.001)	22.67 (<0.001)	13.82 (0.2433)	0.64 (0.4248)	0.88 (0.3475)	0.0006 (0.9803)
May 2008 vs. Oct. 2009	15.02 (0.1818)	45.77 (<0.001)	0.86 (0.3529)	7.11 (0.0077)	16.49 (0.1239)	2.93 (0.0868)	7.88 (0.0050)	3.81 (0.0510)

feeding preference by voles. Ash was the combination of inorganic compounds in food (AOAC 2000) that could be critical for voles (Dubay *et al.*, 2008). Other chemical contents we measured or calculated, including hemicellulous and water did not affect the preference of voles much, although Ho (2009), after performing palatability trials of 13 common meadow plants over a year, found hemicellulous an important nutrient content of plants influencing plant palatability to Taiwan voles. We did not measure crude lipids, secondary compounds, and energy contents of Yushan cane parts in this study, because Ho (2009) found they were very low in amounts, and did not have any effect on palatability.

Although it's nearly impossible to directly observe the consumption behaviors of voles in the field, my feeding experiments suggested that Taiwan voles strongly preferred Yushan cane leafs almost year round. In the shooting season, however, the first-year shoots would be heavily depredated. As shown in Table 11.3, Yushan canes had relatively stable nutrient contents year round except the shooting season when it allocated much energy to emerging new shoots. Taiwan voles are small mammals with high metabolic rates that do not hibernate, and they could reach high densities in some years. Thus, they could greatly depress photosynthesis structure, *i.e.*, leaves, as well as asexual reproduction, *i.e.*, new shoots, of Yushan canes, and potentially affect their fitness negatively. Based on the extent of defoliation we observed during the bite trials, we believe that Taiwan voles could defoliate Yushan canes substantially in the field. Particularly, the cost of searching for high quality leaves should be quite low year round. In the leaf and shoot growing seasons, leaves were all green. In the non-growing season, only the outer layers of foliages particularly the canopy that exposed to wind would wither, while the substratum of foliage remained green (Chen 1997). The fresh clipping piles of leaves and twigs could be found year round in the meadow.

Although at the first glance Taiwan voles seem to benefit from Yushan canes at the cost of the latter, we found the relationship between the two species was more complicated than that. The defoliation of Yushan canes reduces the canopy cover and increases the amount of leaf litter on the ground. The former would increase light penetration to the ground, whereas the latter would maintain higher temperature and humidity on the ground than otherwise. These changes in microclimate could facilitate the emergence of new shoots (Wang and Kao 1986, Liu *et al.*, 2009), thus asexual reproduction of Yushan canes. In deed, both reducing canopy cover and maintaining ground litter had a significant effect on the shoot-culm ratios of Yushan canes. Although the former had a greater effect than the latter one, the shoot-culm ratios increased over time under both treatments (Figure 11.10). The vole exclusion experiment provided further supports. Exclosures that allow vole access had significant higher shoot-culm ratios of Yushan canes than the vole-proof exclosures over a 22 months period. The activity of voles must have altered the growth of Yushan canes. So, the overall effects of Taiwan voles on Yushan canes seemed to be positive, that Yushan canes might have compensatory growth after voles' foraging. Similarly, several studies showed that the debarking of willow stems by voles and lemmings could result in the mortality of damaged stems, while stimulating the emergence of large numbers of new shoots at tree bases (Elmqvist *et al.*, 1987, Predavec and Danell 2001).

Although Taiwan voles loved newly emerging shoots in the laboratory feeding trials, we think the voles might not consume new shoots in the field, as supported by my survey of 99 shoots. Only 8 out of 99 first-year shoots in a 25 meter square area had evidence of vole herbivory after a year. The result was surprising given voles' love for new shoots in the laboratory, and the high vole density at the survey area (Ho 2009). However, it was likely that voles intentionally avoided consuming new shoots in the field to increase the growths of Yushan canes to provide sufficient high quality food year round. Because Taiwan vole presence increased overall shoot-culm ratios of Yushan canes, the latter seemed to benefit from the former. If one accepted that the ends justified the means, then maybe we could say the two species demonstrate a mutualistic relationship.

In conclusion, Taiwan voles did have feeding preference on different parts of Yushan canes in different season. Leaves were preferred year round except in shooting seasons when newly emerged shoots were preferred. Feeding preference can be explained by nutrient contents of different parts: highly preferred parts have lower fiber (ADF) and higher protein (CP) contents than less preferred parts. Consumption by Taiwan voles increased the shoot/culm ratios and could have overall positive effects on the growth of Yushan canes. We suggest that the two species demonstrate a mutualistic relationship.

References

Allison, P. D. (2005). Logistic regression using SAS : theory and application. Wiley, [New York].

Alonso-Diaz, M. A., Torres-Acosta, U. J., Sandoval-Castro, C. A., Hoste, H., Aguilar-Caballer, A. J., and Capetillo-Leal, C. M. (2008). Is goats' preference of forage trees affected by their tannin or fiber content when offered in cafeteria experiments? Animal Feed Science and Technology **141**:36-48.

AOAC. (2000). Official Method of Analysis. 18 edition. Associations of Official Analytical Chemists., Arlington, Virginia. USA.

Bergeron, J. M., and Jodoin, L. (1987). Defining High-Quality Food Resources of Herbivores - the Case for Meadow Voles (Microtus-Pennsylvanicus). Oecologia **71**:510-517.

Borghi, C. E., and Giannoni, S. M. (1997). Dispersal of geophytes by mole-voles in the Spanish Pyrenees. Journal of Mammalogy **78**:550-555.

Chang, M. H. (1981). Ecology and Control of Yushan Cane. National Taiwan University Master Thesis.

Chen, Y. F. (1997). Taiwan Vegetation Vol.3 : Subapline Coniferous Forest and Alpine Meadow. Avanguard Press, Taipei.

Darabant, A., Rai, P. B., Tenzin, K., Roder, W., and Gratzer, G. (2007). Cattle grazing facilitates tree regeneration in a conifer forest with palatable bamboo understory. Forest Ecology and Management **252**:73-83.

Deguchi, Y., Sato, S., and Sugawara, K. (2001). Relationship between some chemical components of herbage, dietary preference and fresh herbage intake rate by the Japanese serow. Applied Animal Behaviour Science **73**:69-79.

Derting, T. L., and Hornung, C. A. (2003). Energy demand, diet quality, and central processing organs in wild white-footed mice (Peromyscus leucopus). Journal of Mammalogy **84**:1381-1398.

Dubay, S. A., Hayward, G. D., and del Rio, C. M. (2008). Nutritional value and diet preference of arboreal lichens and hypogeous fungi for small mammals in the Rocky Mountains. Canadian Journal of Zoology-Revue Canadienne De Zoologie **86**:851-862.

Elmqvist, T., Ericson, L., Danell, K., and Salomonson, A. (1987). Flowering, Shoot Production, and Vole Bark Herbivory in a Boreal Willow. Ecology **68**:1623-1629.

Feeley, K. J., and Terborgh, J. W. (2005). The effects of herbivore density on soil nutrients and tree growth in tropical forest fragments. Ecology **86**:116-124.

Goering, H. K., and Vansoest, P. J. (1970). Forage fiber analyses. Agriculture Handbook No.379. United States Department of Agriculture, Washington D.C., USA.

Goldberg, M., Tabroff, N. R., and Tamarin, R. H. (1980). Nutrient Variation in Beach Grass in Relation to Beach Vole Feeding. Ecology **61**:1029-1033.

Gomez-Garcia, D., Azorin, J., Giannoni, S. M., and Borghi, C. E. (2004). How does Merendera montana (L.) Lange (Liliaceae) benefit from being consumed by mole-voles? Plant Ecology **172**:173-181.

Gomez-Garcia, D., Giannoni, S. M., Reine, R., and E. Borghi, C. (1999). Movement of seeds by the burrowing activity of mole-voles on disturbed soil mounds in the Spanish pyrenees. Arctic Antarctic and Alpine Research **31**:407-411.

Gough, L., Shrestha, K., Johnson, D. R., and Moon, B. (2008). Long-term mammalian herbivory and nutrient addition alter lichen community structure in Alaskan dry heath tundra. Arctic Antarctic and Alpine Research **40**:65-73.

Harju, A., and Hakkarainen, O. (1997). Effect of protein and birch-bark powder on selection of food by root voles (Microtus oeconomus). Journal of Mammalogy **78**:563-568.

Ho, H. C. (2009). The interaction between Taiwan Vole (Microtus kikuchii) and alpine meadow herbs at the Ho-huan area: effect of nutrition and abundance of dominant herbs. National Taiwan University Master Thesis.

Hudson, J. M. G., Morrison, S. F., and Hik, D. S. (2008). Effects of leaf size on forage selection by collared pikas, Ochotona collaris. Arctic Antarctic and Alpine Research **40**:481-486.

Hummel, G. M., Schurr, U., Baldwin, I. T., and Walter, A. (2009). Herbivore-induced jasmonic acid bursts in leaves of Nicotiana attenuata mediate short-term reductions in root growth. Plant Cell and Environment **32**:134-143.

Huntly, N. (1991). Herbivores and the Dynamics of Communities and Ecosystems. Annual Review of Ecology and Systematics **22**:477-503.

Krebs, C. J. (1999). Ecological Methodology. Menlo Park: Addison Wesley Longman Inc.

Liao, M. C. (2004). Studies on Reproduction Biology of Yushania niitakayamensis (Hay.) Keng f. National Chung Hsing University Master Thesis.

Lin, C. Y., and Lin, L. K. (1989). The survery of alpine meadow ecosystem in Taroko National Park. Taroko National Park, HuaLien.

Lin, W. C. (1976). The Classification of Subfamily Bambusoideae in Taiwan (continued). Bulletin of the Taiwan forestry research institute **271**.

Liu, K. C., Yen, S. H., Yeh, C. Y., Li, L. S., and Chang, T. R. (2009). The SOP of Bambusa oldhamii Munro production and management. Agriculture and Food Agency, Council of Agriculture, Executive Yuan, Taipei.

Morrison, S. F. and Hik,. D. S. (2008). Discrimination of intra- and inter-specific forage quality by collared pikas (Ochotona collaris). Canadian Journal of Zoology-Revue Canadienne De Zoologie **86**:456-461.

Parsons, M. H., Lamont, B. B., Davies, S. J. J. F., and Kovacs, B. R. (2006). How energy and coavailable foods affect forage selection by the western grey kangaroo. Animal Behaviour **71**:765-772.

Predavec, M., and Danell, K. (2001). The role of lemming herbivory in the sex ratio and shoot demography of willow populations. Oikos **92**:459-466.

Ritchie, M. E., Tilman, D., and Knops, J. M. H. (1998). Herbivore effects on plant and nitrogen dynamics in oak savanna. Ecology **79**:165-177.

Rosumek, F. B., Silveira, F. A. O., Neves, F. D., Barbosa, N. P. D., Diniz, L., Oki, Y., Pezzini, F., Fernandes, G. W., and Cornelissen, T. (2009). Ants on plants: a meta-analysis of the role of ants as plant biotic defenses. Oecologia **160**:537-549.

SAS. (2003). SAS 9 for Windows. NC: Statistical Analysis Systems Institute Inc., Cary.

Shen, M. L. (2005). Experimental Designs. Jeou Chou Book Co,Ltd., Taipei.

Stuart-Hill, G. C. and Mentis, M. T. (1982). Coevolution of African grasses and large herbivores. African Journal of Range and Forage Science **17**:122-128.

Tripathi, S. K. and Singh, K. P. (1994). Productivity and Nutrient Cycling in Recently Harvested and Mature Bamboo Savannas in the Dry Tropics. Journal of Applied Ecology **31**:109-124.

Vansoest, P. J., Robertson, J. B., and Lewis, B. A. (1991). Methods for Dietary Fiber, Neutral Detergent Fiber, and Nonstarch Polysaccharides in Relation to Animal Nutrition. Journal of Dairy Science **74**:3583-3597.

Wagner, D. (1997). The influence of ant nests on Acacia seed production, herbivory and soil nutrients. Journal of Ecology **85**:83-93.

Wang, T. D., and Kao, Y. B. (1986). The growth and development of bamboo. Modern silviculture.

Wang, W., Franklin, S. B., and Cirtain, M. C. (2007a). Seed germination and seedling growth in the arrow bamboo Fargesia qinlingensis. Ecological Research **22**:467-474.

Wang, W., Franklin, S. B., and Ouellette, J. R. (2007b). Clonal regeneration of an arrow bamboo, Fargesia qinlingensis, following giant panda herbivory. Plant Ecology **192**:97-106.

Widmer, Y. (1998). Pattern and performance of understory bamboos (Chusquea spp.) under different canopy closures in old-growth oak forests in Costa Rica. Biotropica **30**:400-415.

Willig, M. R. and Lacher, T. E. (1991). Food Selection of a Tropical Mammalian Folivore in Relation to Leaf-Nutrient Content. Journal of Mammalogy **72**:314-321.

Wu, J. S. (2007). Mating System of Taiwan Vole (Microtus kikuchii): Evidence From Field Data and Microsatellite DNA. Tung Hai University Master Thesis.

Zavada, M. S., and Mentis, M. T. (1992). Plant-Animal Interaction - the Effect of Permian Megaherbivores on the Glossopterid Flora. American Midland Naturalist **127**:1-12.

Chapter 12

Helminth Parasitism in Rodents in Bamboo Growing Areas of Mizoram: Potential Candidates for Biological Control

Veena Tandon and C. Malsawmtluangi

Department of Zoology, North-Eastern Hill University, Shillong – 793 022, India

Introduction

Rodents are one of the most abundant and destructive pests inflicting incalculable losses to standing crops, harvested crops in threshing floors, stored food grains and other commodities (Figure 12.1). They act as host to many species of helminth parasites, which can be transmitted to human being and other vertebrates. Rodents also serve as reservoir host and aid in dissemination of several parasitic infections to domestic animals and man, thus causing zoonoses.

In the state of Mizoram (Northeast India), bamboo forests account for and cover about 29 per cent of the total forest area, and comprise more than 20 species (representing 9 genera) of bamboo. Of these, *Melocanna* is the most predominant species, contributing about 95 per cent of the growing stock of bamboo. Most bamboo species characteristically show a periodicity of flowering and flower at a species-specific time interval. For example, *Melocanna baccifera* flowers periodically, at an interval of 48±1 years. The phenomenon of periodic flowering is locally known as 'Mautam'; its latest occurrence was reported in the year 2007 in Mizoram. This event of bamboo

Figure 12.1: Rodents, the most Abundant and Destructive Pests of bamboo.

flowering is intertwined with rodent outbreaks in the region, which are believed to cause tremendous destruction to food crops and consequently result in famine. There is also evidence to suggest that occurrence of famine in concurrence with the bamboo flowering phenomemon is a real happening and not merely a superstitious belief (John and Nadgauda, 2002).

Spectrum and Prevalence of Helminth Parasites in Rodents

A variety of parasite species are known to infect rodents worldwide. Among these, helminths (including trematode, cestode, nematode and acanthocephala or spiny-headed worms) constitute a major group of parasites. The parasite spectrum in rodent hosts comprises about 58 species of trematodes (representing 20 families), more than 25 species of cestodes (under 5 families), more than 40 species of nematodes (under 15 families) and only one species of Acanthocephala (family Moniliformidae) reported so far from across the globe. However, in India this spectrum is much narrower and is represented by 8 species under 6 genera of trematodes, 11 species and 6 genera of cestodes, 17 species under 11 genera of nematodes and only one species of acanthocephala that have been reported from various species of rodents.

Although various measures have been taken to control rodents, exhaustive studies on the parasite fauna, which can be harmful to rodents, have never been undertaken. In view of the underlying threat posed by rodents as serious pests of crop plants and also as reservoir of zoonoses, a study on the parasitic infections prevailing in rodent hosts in Mizoram was undertaken. A number of parasites belonging to the helminthic group were found to parasitize the rodent hosts. A total of 280 rodents were collected from 10 different locations (Table 12.1) and screened for the presence of helminth parasites (Figure 12.2), the commonly prevalent rodent species in the region being *Rattus rattus, R. nitidus, R. norvegicus, Bandicota bengalensis, Berylmys mackenziei, B. bowersi, Mus musculus, Niviventer fulvescens* and *Cannomys badius*. The parasite species recovered along with their distribution, prevalence, intensity and abundance are listed in Tables 12.2 and 12.3.

Of the 280 rodents surveyed representing 6 genera and 9 species collected from 10 different locations, *Rattus rattus* showed a wide range of parasites with 8 nematode and 4 cestode species; *Rattus nitidus* with 8 nematode, 3

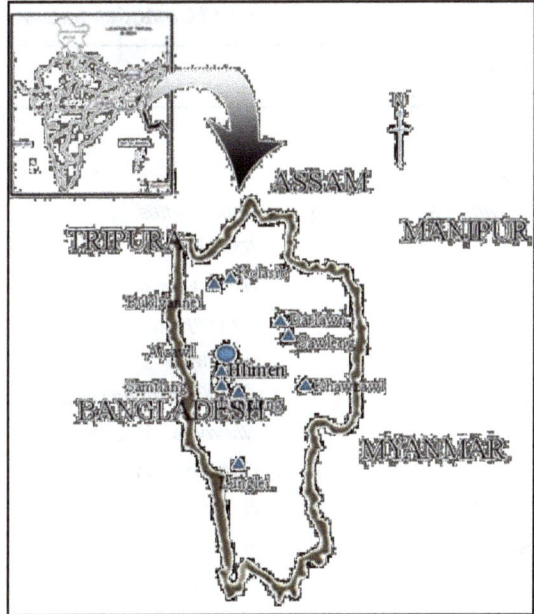

Figure 12.2: Rodent collection sites from Mizoram.

cestode and 1 acanthocephalan species; both collected from nearby human dwellings (Table 12.4). *Cannomys badius* showed the lowest range of parasite infection. Nematodes and cestodes emerged as the parasite groups dominating the helminth fauna of rodents. The most prevalent and abundant species were *Hymenolepis diminuta* and the metacestode *Cysticercus fasciolaris* among the cestodes and *Capillaria hepatica* among the nematodes. The conspicuous absence of any trematode infection in rodents of the study area may be attributed to the fact that the trematode life cycle involves implication of aquatic intermediate hosts, *i.e.*, snails.

Table 12.1: Collection sites with their geographical locations.

Sl.No.	Collection Site	Latitude	Longitude	Altitude
1.	Kolasib	24° 13" 25'N	92° 40" 39'E	2066 ft
2.	Bukvannei	24°11" 49' N	92° 32" 15' E	184 ft
3.	Hlimen	23° 42" 04' N	92° 43" 01' E	3433 ft
4.	Samtlang	23° 41" 43' N	92° 43" 07' E	3173 ft
5.	Lungleng	23° 41" 01' N	92° 42" 55' E	3415 ft
6.	Aizawl	23° 43" 38' N	92° 44" 11' E	2389 ft
7.	Khawzawl	23° 31" 14' N	93° 11" 15' E	3971 ft
8.	Darlawn	23° 58" 48' N	92° 59" 34' E	955 ft
9.	Sawleng	23° 54" 20' N	92° 59" 20' E	2139 ft
10.	Lunglei	22° 57" 27' N	92° 48" 54' E	2914 ft

Table 12.2: Distribution of Cestodes and Acanthocephala in Rodent Hosts.

Group and name of parasite	Hosts	Prevalence (Per cent)	Mean Intensity	Abundance
Cestoda:				
Cyclophyllidea:				
Hymenolepis diminuta	*Rattus rattus*	31.8	4.64	1.47
	R. nitidus	15.6	4	0.62
	R. norvegicus	**60**	**9.1**	**5.5**
	Berylmys mackenziei	36.37	2.8	1.04
	B. bowersi	33.34	1.84	0.94
	Bandicota bengalensis	25	3.2	0.8
	Niviventer fulvescens	26.67	7.25	1.94
	Cannomys badius	20	5	1
Rodentolepis sp.	*R. rattus*	2.28	**3.5**	0.07
	R. norvegicus	**30**	3	0.3
	B. bengalensis	10	2.5	0.25
	N. fulvescens	6.67	7	**0.47**
	Mus musculus	2.28	5	0.11
Raillietina celebensis	*R. rattus*	2.27	1.5	0.03
	R. nitidus	3.12	**3**	0.09
	B. bengalensis	5	2	0.1
	N. fulvescens	**6.67**	2	**0.13**
Taeniidae:				
Cysticercus fasciolaris	*R. rattus*	29.5	1.3	0.38
	R. nitidus	18.75	**3.34**	0.62
	R. norvegicus	40	1.75	**0.7**
	Berylmys mackenziei	**40.9**	1.34	0.54
	B. bowersi	16.67	1	0.16
	B. bengalensis	30	1.16	0.35
	N. fulvescens	6.67	1	0.06
	M. musculus	4.54	1	0.04
Acanthocephala				
Moniliformis moniliformis	*R. nitidus*	1.56	15	0.23

A number of new hosts were recorded for several species of parasites as summarized below:

1. **R. nitidus**- *Raillietina celebensis, Moniliformis moniliformis, Cysticercus fasciolaris, Trichuris muris, Capillaria hepatica, Trichosomoides crassicauda, Heterakis spumosa, Rictularia* sp., *Syphcia obvelata* and *Nippostrongylus brasiliensis.*

Table 12.3: Distribution of Nematodes in Todent Hosts.

Group and name of parasite	Hosts	Prevalence (Per cent)	Mean Intensity	Abundance
Nematoda:Adenophorea: **Trichinelloidea:**				
Trichuris muris	*R. rattus*	1.13	1	0.01
	R. nitidus	**6.25**	**2.5**	**0.15**
Capillaria hepatica	*R. rattus*	29.54	–	–
	R. nitidus	**40.1**	–	–
	B. mackenziei	31.8	–	–
	B. bowersi	16.6	–	–
	N. fulvescens	40	–	–
	M. musculus	2.27	–	–
Trichosomoides crassicauda	*R. rattus*	5.68	3.2	0.18
	R. nitidus	1.56	1	0.01
	R. norvegicus	10	3	0.3
	B. mackenziei	9.09	1	0.09
	B. bowersi	**16.6**	1.5	0.25
	B. bengalensis	15	**6.66**	**1**
	N. fulvescens	6.67	1	0.06
Nematoda:Secernentea: **Heterakoidea:**				
Heterakis spumosa	*R. rattus*	7.95	4.85	0.38
	R. nitidus	7.81	2.8	0.21
	B. mackenziei	4.54	2	0.09
	B. bengalensis	**25**	**17.8**	**4.45**
Rictularioidea:				
Rictularia sp.	*R. rattus*	**4.49**	1	0.04
	R. nitidus	3.12	**5**	**0.15**
Oxyuroidea:				
Syphacia obvelata	*R. rattus*	7.95	6.1	0.48
	R. nitidus	**9.37**	4.67	0.43
	B. mackenziei	4.54	7	0.31
	M. musculus	9.09	**13.5**	**1.2**
Aspicularis (Paraspicularis) pakistanica	*B. mackenziei*	**4.54**	3	0.13
	M. musculus	2.27	**56**	**1.27**
Trichostrongyloidea:				
Nippostrongylus brasiliensis	*R. rattus*	17.04	27.14	4.62
	R. nitidus	**23.43**	17	3.98

Contd...

Table 12.3–*Contd...*

Group and name of parasite	Hosts	Prevalence (Per cent)	Mean Intensity	Abundance
	R. norvegicus	10	5	0.5
	B. mackenziei	18.18	10.5	1.9
	B. bowersi	8.34	10	0.84
	B. bengalensis	5	**45**	2.25
	M. musculus	15.9	31	**4.93**
Hepatojarakus bandicoti	*R. rattus*	**7.95**	1.57	0.12
	R. nitidus	7.81	**1.8**	**0.14**
	B. mackenziei	4.54	1	0.04
	B. bengalensis	5	1	0.05
	M. musculus	2.27	1	0.02
	N. fulvescens	6.67	1	0.06

Table 12.4: Distribution of helminth species in rodents host.

Host	*Parasite*		
	Cestode	*Nematode*	*Acanthocephala*
R. rattus	4	8	–
R. nitidus	3	8	1
R. norvegicus	3	2	–
B. mackenziei	2	7	–
B. bowersi	2	3	–
B. bengalensis	4	4	–
M. musculus	2	5	–
N. fulvescens	4	3	–
Cannomys badius	1	–	–

2. ***B.bowersi-*** *Hymenolepis diminuta, Cysticercus fasciolaris, Capillaria hepatica, Trichosomoides crassicauda,* and *Nippostrongylus brasiliensis.*

3. ***B. mackenziei-*** *Hepatojarakus bandicoti, Nippostrongylus brasiliensis, Aspiculuris (Paraspiculuris) pakistanica, Heterakis spumosa, Trichosomoides crassicauda, Cysticercus fasciolaris* and *Hymenolepis diminuta.*

4. ***N.fulvescens-*** *Hymenolepis diminuta, Rodentolepis sp., Cysticercus fasciolaris, Raillietina celebensis, Capillaria hepatica, Trichosomoides crassicauda* and *Hepatojarakus bandicota.*

5. ***C. badius-*** *Hymenolepis diminuta.*

Molecular Characterization

In order to ascertain the species of the metacestode, molecular characterization of the same was done to supplement the morphological findings. Total genomic DNA was isolated from the recovered metacestode. Ribosomal internal transcribed spacer (ITS) and mitochondrial cytochrome oxidase subunit 1 (CO1) regions were PCR-amplified from the genomic DNA using trematode universal primers (Bowles *et al.*, 1995). The PCR product obtained was sequenced and sequences were submitted to GenBank and their accession numbers obtained (FJ939133, FJ939134 and FJ939135).

PCR amplification of the metacestode DNA yielded the product size of ~450bp, ~750bp and ~350bp in case of ITS1, ITS2 and CO1 regions respectively. Multiple sequence alignment was done for comparison with other available sequences in the database. From the analysis, it was concluded that the metacestode found in the liver of rodents is indeed *Taenia taeniaeformis*.

Histopathology

Two liver parasites- *Cysticercus fasciolaris* (metacestode of *Taenia taeniaeformis*) and *Capillaria hepatica* (nematode) were amongst the most frequently encountered infections in the survey. In view of their frequent occurrence, we also tried to find out the effect of the metacestode and *C. hepatica* on the liver of the infected hosts in order to ascertain their potential for biological control of rodents. For the purpose a histopathological approach was adopted.

Taenia taeniaeformis is a taeniid cestode found in the intestine of cats, other felines and carnivores that serve as definitive hosts. Rodents serve as the intermediate host, in which the larval form or the metacestode, *Cysticercus fasciolaris*, develops in the liver and other organs as a fluid-filled bladder worm (Figure 12.3). Sarcomas of the rat liver due to the presence of *C. fasciolaris* have long been known and several experiments using this metacestode to induce malignant growth of the connective tissues in the liver of rats were successfully conducted as early as the first quarter of the twentieth century (Bullock and Curtis, 1920, 1924-1926, 1928). Liver fibrosarcoma due to the presence of *C. fasciolaris* was suggested to be an appropriate model for studying parasitic carcinogenesis and pathogenesis in wild rats (Tucek *et al.*, 1973).

Capillaria hepatica is a nematode found in the liver of rodents (Figure 12.4) and other lagomorphs that can also parasitize man (Berger *et al.*, 1990; Choe *et al.*, 1993) and has been the most frequently encountered species in wild and house rodents (Junker *et al.*, 1998; Seong *et al.*, 1998). A higher prevalence of this parasite in wild rats was observed in northern parts of India (Mittal, 1980; Gupta and Trivedi 1988; Somvanshi *et al.*, 1995; Chahota *et al.*, 1997). The female worms of *C. hepatica* die soon after laying eggs and disintegrate inside the liver, forming focal necro-inflammmatory lesions that heal by encapsulation, calcification and resorption (Luttermoser, 1938; Ferreira and Andrade, 1993; Gotardo *et al.*, 2000). Eggs are released only when the host dies and its liver decays or when the infected rodent is eaten by another carnivore; in such cases the egg are released with the faeces of the carnivore. Several studies have been done on the pathogenesis of *C. hepatica* and its effect on the liver tissue. In

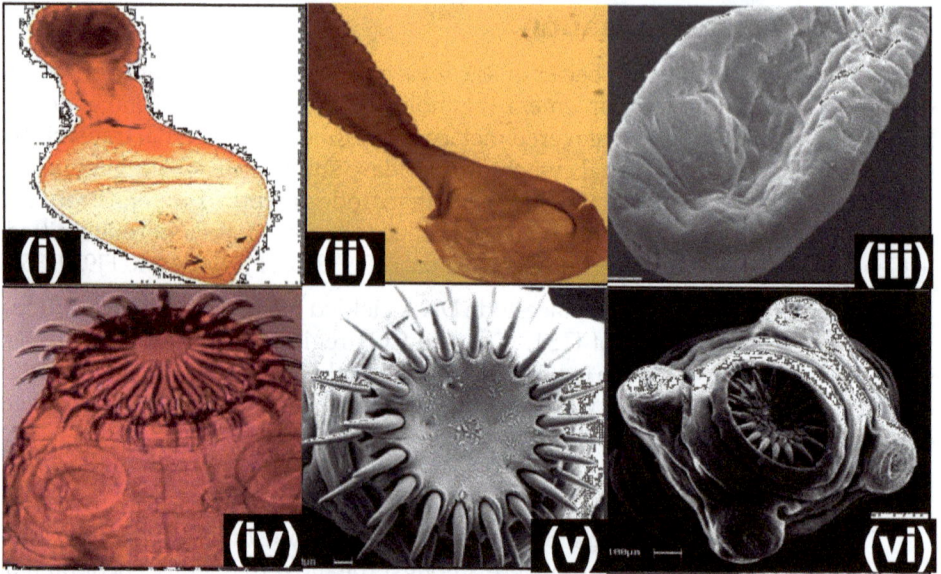

Figure 12.3: Metacestode - *Cysticercus fasciolaris* **in liver tissue.**
(i) Whole metacestode; (ii-iii) Portions of the bladderworm body; (iv-vi) The scolex end showing suckers and armature of typically taenid hooks.

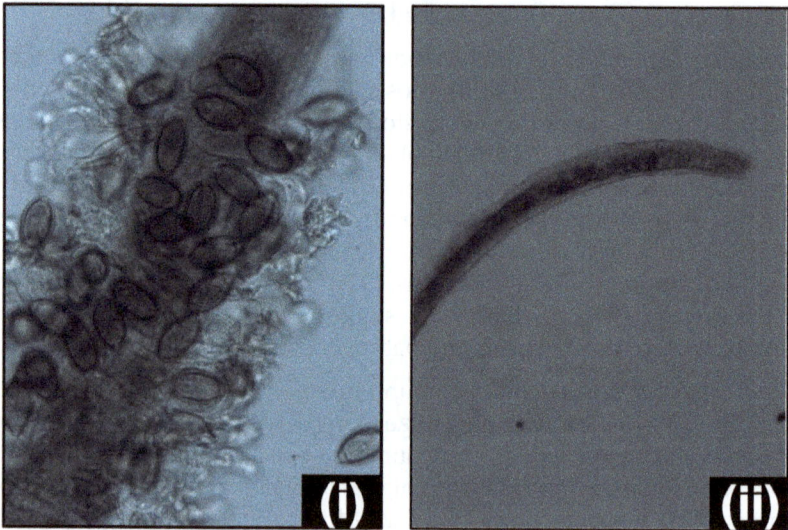

Figure 12.4: *Capillaria hepatica*
(i) Eggs with characteristic bipolar plugs; (ii) Slender worm embedded in liver tissue.

laboratory mice, the infection can reduce the reproductive output or even cause death of the host (Singleton and Spratt, 1986).

Upon necropsy of the rodent host, cream-coloured cysts were observed on the hepatic parenchyma and on opening the cyst, a viable creamy white larva was

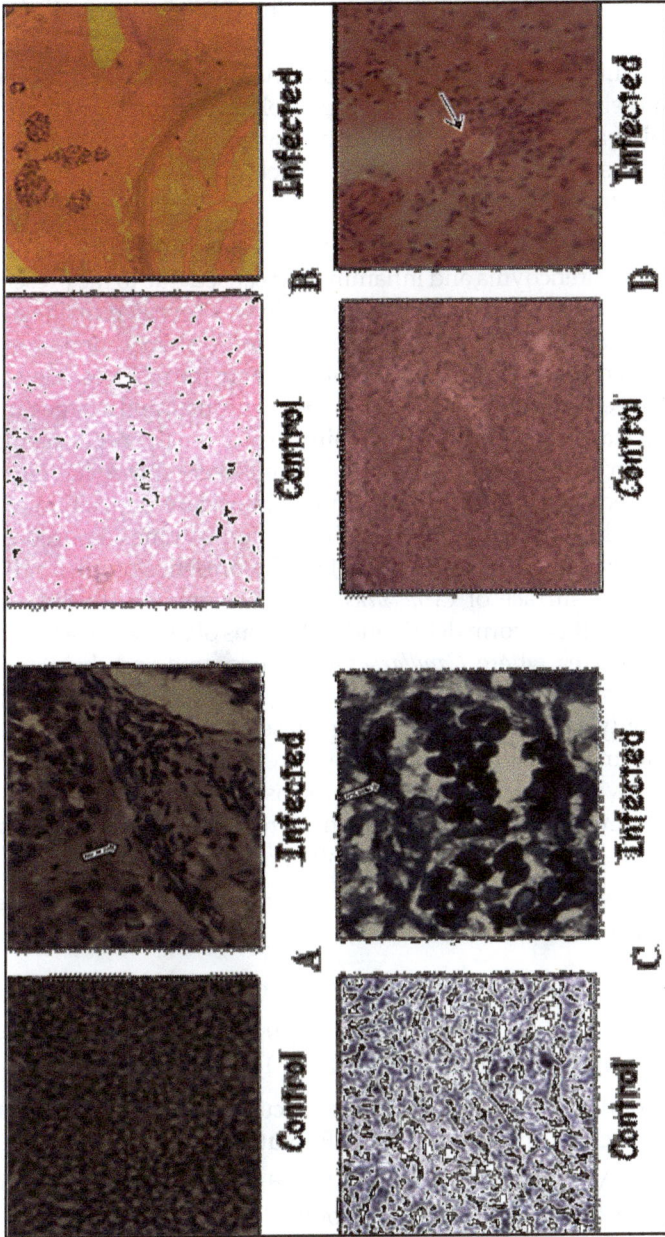

Figure 12.5: Histopathological Observations of the Infected Liver.

A. Infected liver tissue showing spindle shaped cells aggregating near the area where the parasite occurs. B. Haemotoxylin and eosin stain showing fibrous tissue encapsulation of the metacestode and abundant eisonphilic cytoplasm around the area where both the parasites occur adjacently in the infected tissue. C. Presence of lipids in the liver tissue infected with *C. hepatica*. D. Irregular white/ yellowish nodules containing eggs or adult worms scattered on the surface of liver infected with *C. hepatica*.

observed. Presence of *C. hepatica* was also observed by gross examination of the liver tissue and the nematode was easily recognized externally by the presence of irregular white or yellowish white nodules containing the eggs or adult worms and scattered all over the surface of the liver. The liver tissue containing the two parasites was separated and washed with Phosphate Buffered Saline (PBS) and stored at -40 °C. The fresh frozen tissues of the infected and control (uninfected) liver were sectioned in a cryostat (Model no LEICA CM 1850) at 14μm thickness and at -200 °C and stained with hematoxylin and eosin (H and E). For histopathological studies, alterations, if any, in collagen, lipids and number of eosinophils were used as parameters, for detection of which Masson's trichrome, Sudan black and Congo red methods, respectively were used following Pearse (1968).

Histopathological studies of the infected liver revealed distortion of the normal morphology of the liver parenchyma and inflammation caused due to the presence of both the parasite species. The presence of metacestode was revealed inside a well-defined fibrous tissue capsule. The cells appeared spindle shaped and clustered together with abnormal nuclei in and around the area where the metacestode occurs; in some areas the cells seemed to be larger as compared to normal; and a large number of cells were multinucleated and the normal architecture of the liver cells seemed to be altered. With Masson's trichrome stain neoplastic cells were shown having black coloured nuclei with red coloured cytoplasm and an abundant deposition of blue coloured collagen sheath. With Sudan black stain, numerous blue coloured nuclei were found in the *C. hepatica* - infected liver; in fact the presence of the lipids was found to be more on the surface of *C. hepatica* eggs as compared to the hepatic parenchyma. Partially calcified worm debris and collections of immature and mature eggs were found in the area where *Capillaria* worms occurred and disintegrated. Granulomatous lesions surrounding the nematode eggs were detected. Sometimes the lesion contained only a calcified core, besides these inflammatory lesions there also occurred septal formations within the infected liver. Clusters of nematode eggs having ovoid structure with bipolar caps were clearly visible on the liver parenchyma. Abundant eosinophilic cytoplasm was observed in the region where the metacestode of *Taenia* species. and *C. hepatica* occurred adjacently (Figure 12.5).

These observations suggest that the metacestode and *Capillaria* could be used as a potential biological control tool of rodents.

References

Berger, T, Degremont, A., Gebbers, J.O. and Tonz, O. (1990). Hepatic capillariasis in a 1-year- old child. European Journal of Pediatrics. 149: 333-336.

Bowles, J., Blair, D. and McManus, D. P. (1995). A molecular phylogeny of the human schistosomes. Molecular Phylogenetics and Evolution, **4**: 103-109.

Bullock, F.D. and Curtis, M.R. (1920). Experimental production of sarcoma of the liver of rats. Proceedings of the New York Pathological Society. n. s. 20: 149-171.

Bullock, F.D. and Curtis, M.R. (1924). A study of the reaction of the tissues of the rats liver to the larvae of *Taenia crassicollis* and the histogenesis of Cysticercus sarcoma. Journal of Cancer Research. 8: 446-481.

Bullock, F.D. and Curtis, M.R. (1926). Further studies of the transplantation of the larvae of Taenia crassicollis and the experimental production of subcutaneous cysticercus sarcomata. Journal of Cancer Research. 10: 393-421.

Bullock, F.D. and Curtis, M.R. (1928). A cysticercus carcino- osteochondrosarcoma of the rat liver with multiple cysticercus sarcomata. Journal of Cancer Research. 12: 326-334.

Chahota, R., Asrani, R.K., Katoch, R.C. and Jitendra, K.P. (1997). Hepatic capillariasis in a wild rat (*Rattus rattus*). Journal of Veterinary Parasitology. 11: 87-90.

Choe, G., Lee, H.S., Seo, J.K., Chai, J.Y., Lee, S.H., Eom, K.S. and Chi, J.G. (1993). Hepatic capillariasis: First case report in the Republic of Korea. American Journal of Tropical Medicine and Hygiene. 48: 610-625.

Ferreira and Andrade (1993). *Capillaria hepatica*: a cause of septal fibrosis in the liver. Memorias do Institudo Oswaldo Cruz. 88: 441-447.

Gotardo, B.M., Andrade, R.G. and Andrade, Z.A.. (2000). Hepatic pathology in *Capillaria hepatica* infected mice. Revista da Sociedade Brasileira de Medicina Tropical. 34: 341-346.

Gupta, S.P. and Trivedi, K.K. (1988). Nematode parasites of vertebrates. On a new spirurid, *Protospirura srivastavai* sp. nov. (Family: Spiruridae Oerley, 1885) from a field mouse, *Mus platythrix* from Udaipur, Rajasthan. Indian Journal of Helminthology. 39: 153-159.

John, C.K. and Nadgauda, R.S. (2002). Bamboo flowering and famine. Current Science, 82:261-262.

Junker, V., Voss, M., Kubber-Heiss, A. and Prosi, H. (1998). Detection of *Capillaria hepatica* in house mice (*mus musculus*) in Austria. Tagung der Osterreichischen Gesellschaft fur Tropenmedizine und Parasitologie. 20: 137-143.

Luttermoser, G.W. (1938). An experimental study of *Capillaria hepatica* in the rat and the mouse. American Journal of Tropical Medicine and Hygiene. 27: 321-340.

Mittal, R.P. (1980). Frequency and intensity of nematodes in rats and mice of Meerut Division, U.P. Indian Journal of Parasitology, 3 (Supplement): pp. 130.

Pearse, A.G.E. (1968). Histochemistry: Theoretical and applied. 3rd edn. Vol. 1, Churchill Livingstone Edinburgh, London, New York.

Seong, J.K. (1998). Spontaneous infection of *capillaria hepatica* in wild rats (*Rattus norvegicus*) of Korea. Korean Journal of Veterinary Research. 38: 600-605.

Singleton, G.R. and Spratt, D.M. (1986). The effects of *Capillaria hepatica* (Nematoda) on natality and survival to weaning in BALB/c mice. Australian Journal of Zoology. 34:677-681.

Somvanshi, R., Bhattacharya, D., Laha, R. and Rangarao, G.S.C. (1995). Spontaneous *Capillaria hepatica* infestation in wild rats (*Rattus rattus*). Indian Journal of Veterinary Pathology, 19: 44-45.

Tucek, P.C., Woodard, J.C. and Moreland, A.F. (1973). Fibrosarcoma associated with *Cysticercus fasciolaris*. Laboratory Animal Science. 23: 401-407.

Index

www.ingramcontent.com/pod-product-compliance
Lightning Source LLC
Chambersburg PA
CBHW050518190326

41458CB00005B/1578